Supercharge Your Images: Super-Resolution for Enhanced Quality

Hazel

Copyright © 2024 by Hazel

All rights reserved. No part of this book may be reproduced in any manner whatsoever without written permission except in the case of brief quotations embodied in critical articles and reviews.
First Printing, 2024

Table of Contents

1 **Introduction** 1

  1.1 Super-Resolution 1

    1.1.1 Single Image Super-Resolution 1

    1.1.2 Video Super-Resolution 2

    1.1.3 Learning-Based Super-Resolution 3

  1.2 Research Problems 4

  1.3 Contributions of the **book** 6

  1.4 Publication Generated from the **book** Research 7

1.5   book Structure   7

2   Related Concepts   9

  2.1   Convolutional Neural Networks   10

    2.1.1   Layers   10

    2.1.2   Activation Function   15

    2.1.3   Loss Function   18

  2.2   CNN Training   19

  2.3   CNN Related Networks   25

    2.3.1   Multi-Kernel CNN   26

    2.3.2   ResNet   28

    2.3.3   Dilated Convolution   29

    2.3.4   Deformable Convolution   30

    2.3.5   Spatial Transformation Network   32

  2.4   Attention Mechanism   34

    2.4.1   Channel Attention Mechanism   34

    2.4.2   Spatial and Temporal Attention Mechanisms   36

3   Literature Review   38

  3.1   Single Image Super-Resolution   38

  3.2   Video Super-Resolution   44

  3.3   Summary   49

4   Video Super-Resolution via Multi-Kernel Deformable 3D Convolution .   50

  4.1   Framework of Model   52

  4.2   3D Multi-Kernel Attention Network   54

  4.3   Feature Fusion with Multi-Kernel Deformable Convolution   58

  4.4   2D Multi-kernel Attention Network   60

  4.5   Reconstruction Network   61

4.6   Adjustable Multi-Kernel Structure 63

4.7   Summary   65

5   Experimental Results 67

5.1   Environment Setup 67

    5.1.1   Training Datasets 67

    5.1.2   Testing Datasets   69

5.2   Training Strategy 70

5.3   Evaluation and Loss Function   70

5.4   Color Space   72

5.5   Experimental Results 74

    5.5.1   Comparison with Other VSR Methods   75

    5.5.2   Ablation Studies   82

    5.5.3   Evaluation on Additional Examples 88

5.6   Summary   93

6   Conclusions 94

References 96

Chapter 1

Introduction

1.1 Super-Resolution

Super-resolution (SR) aims to reconstruct visually realistic high-resolution (HR) images or videos from low-resolution (LR) counterparts. Super-resolution can be applied to image processing fields such as satellites, medical images, and object recognition [21, 22]. According to the number of input frames processed, SR research has two main focuses: single image super-resolution (SISR) and video super-resolution (VSR).

1.1.1 Single Image Super-Resolution

The goal of SISR is to estimate high-resolution images from a single image. The easiest way to improve image resolution is interpolation [23], but the disadvantage of this method is low visual quality. A single image's interpolation cannot recover the high-frequency part lost in the LR sampling process, resulting in poor restoration in sharp edges. To provide high quality high-resolution images, SISR methods utilize the internal similarity of the images to perform image restoration. Most recent SISR methods learn the mapping between the LR and HR images for restoration using data sets containing low-resolution images and high-resolution images.

1.1.2 Video Super-Resolution

The goal of VSR is to comprehensively process multiple frames of images to restore high-resolution videos. Different from SISR, the reconstruction process of video super-resolution (VSR) benefits from the fusion of multiple consecutive frames, which makes the VSR more potential to obtain higher quality results than SISR. Fig. 1.1 shows the comparison results of the same model with a single image as input and continuous frames as input. It can be observed that the effective use of multiple consecutive frames can significantly improve the quality of HR images.

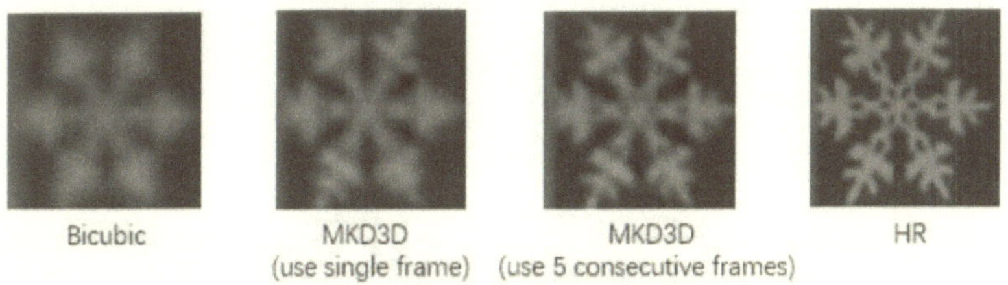

Fig. 1.1 Performance comparison of SISR and VSR.

The motion of the objects in the video results in different scales, different locations, and different blurs. Whether the same object in different frames can be effectively aligned has become the factor that most affects the performance of the VSR model [24]. Therefore, most VSR methods are divided into two similar modules: a motion compensation module and a super-resolution module [18]. Early VSR methods process motion compensation for multi-frame alignment as preprocessing stages, such as optical flow motion estimation methods.

Many studies have shared weights between motion estimation and super-resolution modules to build end-to-end models in recent years, and these studies achieved acceptable performance. Another factor that affects the effect of VSR is the frame rate of the video. In data-based VSR methods, if the frame rate of the training set and the test set do not match, it will cause significant performance degradation. How to adapt the model to different frame rates is also a concern. Finally, how to effectively fuse the information of each frame is a critical step. Before learning-based methods were widely applied to VSR, the fusion of time and space information limited the progress of VSR. In order to solve these problems, learning methods have become the most popular choice for VSR.

1.1.3 Learning-Based Super-Resolution

The breakthrough development of CNN has promoted the rapid progress of SR technology. As a result, SR technology has received extensive attention in recent years. Following SRCNN [8], which is considered the pioneering work of learning-based SISR methods, SR methods based on deep learning demonstrate superior performance [25]. The data set has a significant impact on the learning-based SR method. At present, VSR has a standard test data set, but the lack of a general training data set has led to unfair comparisons between different VSR methods. However, as a result, more public data sets used for SR have been generated. An appropriate training data set has become an essential factor affecting the VSR model's performance.

1.2　Research Problems

SR has a different emphasis in different application fields. The development of the SR model mainly has two directions: faster processing speed and higher restoration quality. With the advancement of hardware, large models can also have faster processing speeds. On the contrary, with the development of electronic display devices, there is a greater demand for high-quality images and videos. Therefore, most SR methods focus on higher restoration quality in recent years [19, 26, 20, 27, 17].

In this **book**, we focus on VSR tasks. Unlike SISR, VSR uses time-bandwidth (acquiring multiple consecutive frames of the same scene) in exchange for spatial resolution. VSR has two main challenges: Firstly, how to align the multiple consecutive frames? Secondly, how to effectively fuse the information of the aligned frames? Existing VSR research utilizes a variety of methods to deal with these two challenges. However, there is still a gap between the video restoration results of the VSR models and the corresponding HR videos. Therefore, VSR still has room for improvement in the quality of video restoration.

Many deep convolutional neural network (CNN) based VSR methods have achieved considerably good performance. These methods imitate the biological visual nervous system, which restores the quality of blurred images by learning from a large amount of data.

Biological visual cortex cells adjust their receptive fields a ccording t o d ifferent inputs [2, 28], which is rarely considered by VSR research. In biological vision, the entire photoreceptors that affect a certain visual cortex cell are called the receptive field o f t he visual cortex cell. Kuffler et al. [29] clarified t hat t he r eceptive fi elds of th e ca t's vi sual cortex cells are concentric circles in the spatial domain. Rodieck et al. [30] proposed a homocentric opponent model of the receptive field. This model contains a central region with a strong stimulus-response and a larger surrounding region with a weak stimulus-response. These two regions antagonize each other to make the visual cortex cell more sensitive to contrast information. By adjusting the strength of this antagonistic effect, the visual cortex can adjust the size of the active field according to the optical input. Also, there are the different responses of the different receptive fields in the visual cortex. These different receptive fields enable the visual cortex to obtain information at various frequencies.

Adaptive receptive fields and diverse receptive fields help the biological vision to process various optical information. It is a collaborative work of the brain and these visual cortex cells to understand blurred images. However, most existing VSR research focuses on training the brain to understand the image process while ignoring the training of visual cortex cells. Specifically, most studies use a fixed-size convolution kernel to construct a 2D or 3D CNN in the same layer, limiting the adaptability of the receptive field.

The primary focus of our research is to improve the reconstruction quality of the VSR. On the premise of not losing the quality of video restoration, we also focus on reducing the computational burden to improve the model's efficiency. We are committed to establishing an end-to-end VSR model to achieve the best video restoration quality on the public test data set.

1.3　Contributions of the **book**

To solve the above research problems, we explore adaptive receptive fields to VSR tasks. This book aims to imitate biological visual cells and constructs an end-to-end VSR network with adaptive receptive fields. The receptive field of a convolutional neural network refers to the region of the input feature map that affects a specific network component. The size of each convolution kernel thus directly affects the receptive field of the network. We deploy multiple convolution kernels to imitate the diverse receptive fields in the visual cortex. Also, we design a weighted fusion mechanism between channels of multiple kernels to imitate adaptive receptive fields of the visual cortex cells. To cope with the challenges of VSR, we deployed the adaptive receptive field in multi-frame alignment and multi-frame information fusion to improve the quality of video restoration. Moreover, we extend the adaptive receptive field to the time domain by proposing a multi-kernel deformable 3D convolution. Our proposed method has three main contributions:

Firstly, we utilize channel attention and separate 3D convolution to create a multi-kernel attention 3D convolution (3D MKA) module. The 3D MKA module can adaptively adjust the receptive field of spatial and temporal dimensions to utilize spatial-temporal information of VSR fully.

Secondly, we propose a multi-kernel deformable convolution (MKDC) for multi-frame alignment. The MKDC module is able to provide more extensive and more flexible receptive fields than standard deformable convolution.

Lastly, we design a multi-kernel 2D convolution (2D MKA) structure to extract high-level features, which is proved to be able to restore more realistic object contours than the structures using standard convolutions.

Thanks to the 3D MKA module, MKDC alignment module, and 2D MKA module, our method achieves the highest test result compared to the previous VSR methods.

1.4 Publication Generated from the book Research

Tianyu Dou, Xiafei Yu, and Jiying Zhao. Multi-Kernel Deformable 3D Convolution for Video Super-Resolution. IEEE International Symposium on Broadband Multimedia Systems and Broadcasting, August 04-06, 2021, Chengdu, China.

1.5 book Structure

This book consists of 6 chapters. In addition to Chapter 1, Chapters 2 through 6 are organized as follows.

In Chapter 2, we outline the basic components and the training process of CNN. And we list some CNN-related networks that inspired our book. We also introduce the attention mechanism used in our book.

In Chapter 3, we review essential deep-learning-based SISR methods and deep-learning-based VSR methods.

In Chapter 4, we describe the framework of our method. And we describe in detail each component of our model. We also provide the adjustable multi-kernel structure of our model under different needs.

In Chapter 5, we first introduce the test environment for our experiment. Then, we provide the results of our experiments. And we compare our experimental results with the previous VSR methods. We also discuss the effectiveness of each component of our method by ablation studies.

In Chapter 6, we conclude our **book**. And we discuss how our method can be improved.

Chapter 2

Related Concepts

Over recent years, deep learning has led to breakthroughs in various research fields such as computer vision, natural language processing, and speech recognition [31, 32]. In image processing, CNN is the most important and most commonly used deep learning algorithm [33].

CNNs have various structures but have similar necessary components [34]. In this chapter, we will introduce these basic components, including layers, activation functions, and loss functions in Sec. 2.1. We will then present CNN's training process in Sec. 2.2. Next, we will list the CNN-related networks that inspired our work in Sec. 2.3. Finally, we will describe the attention mechanisms which are utilized in our work in Sec. 2.4.

2.1 Convolutional Neural Networks

A CNN is a variant of a multi-layer perceptron (MLP) developed by Huber and Wiesel [28] in their early research on cat visual cortices. The visual cortex cells are sensitive to sub-regions of the visual input space. The stimulus region that a visual cortex cell can perceive is called the receptive field. These cells of the visual cortex can be divided into two basic types.. The first type of cells responds most to the edge stimulus pattern in the range of the receptive field, and the second type of cells has a larger receptive and has a degree of spatial invariance [35]. Inspired by these two types of visual cortex cells, our work addresses different receptive fields in the same network layer to construct a neural network close to biological vision.

2.1.1 Layers

CNNs are composed of a large number of layers, and the output of each layer is used as the input of the next layer. The basics of the classic CNN are mainly composed of convolutional layers, pooling layers, and fully connected layers.

Convolutional Layer The purpose of the convolutional layer is to extract different features of the input. Each convolutional layer consists of several convolution units, and multi-layered networks can extract more complex features from low-level features. In image processing, the input of the first convolutional layer is generally the original input image, and the convolutional layer convolves the input to generate a feature map and passes this feature map to the next layer [36]. The shapes of the input images and feature maps are represented as $W \times H \times C$, where W and H denote the width and height, respectively and

C denotes the channel of input images and feature maps. Each convolution operation can be expressed as:

$$f(x) = w * x + b, \tag{2.1}$$

where w is the weight of the convolution kernel, x is the input image, and b is the bias. The purpose of a convolutional neural network is to calculate the optimal weight and bias based on the known input and output. The trainable variables w and b are known as parameters in CNNs. Usually, the number of parameters determines the computational burden of the network.

In addition to the convolution kernel, input, and output, the convolution layer has two important concepts: stride and padding. Stride represents the amplitude of each movement of the filter. Usually, the stride is set to 1 in a CNN. However, the size of the feature map after each convolutional layer is smaller than the input when the stride is set to 1. The feature map after multi-layered convolution will be significantly shrunk. Thus, it is important to keep the size of output the same as input by padding. The padding operation is to pad extra zeros around the border of the input matrics. The zero-padding process is shown in Fig. 2.1.

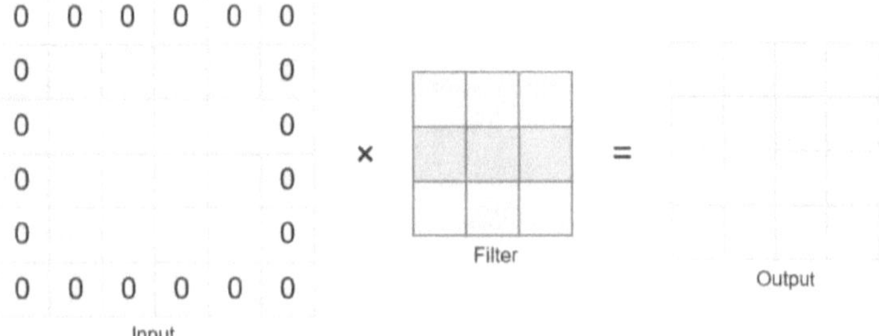

Fig. 2.1 Zero-padding.

Pooling Layer The role of the pooling layer is to downsample and thereby reduce the number of network parameters. Since the downsampling operation of the pooling layer reduces the resolution of the feature map, it should be avoided in the SR models [37]. However, the pooling layer has an irreplaceable role in feature dimensionality reduction. Therefore, the SR models often apply pooling layers in branch networks and use the pooling layer to extract global information without affecting the resolution of the feature map.

Commonly used pooling methods include maximum pooling and average pooling. In the forward process, the maximum pooling calculates the maximum value for each patch of the feature map. In contrast, the average pooling calculates the average value for each patch of the feature map.

Fully Connected Layer Fully connected (FC) layers play the role of "classifiers" in a CNN. Each node of the FC layer is connected to all nodes of the previous layer so that FC layers can integrate the information extracted by the previous network.

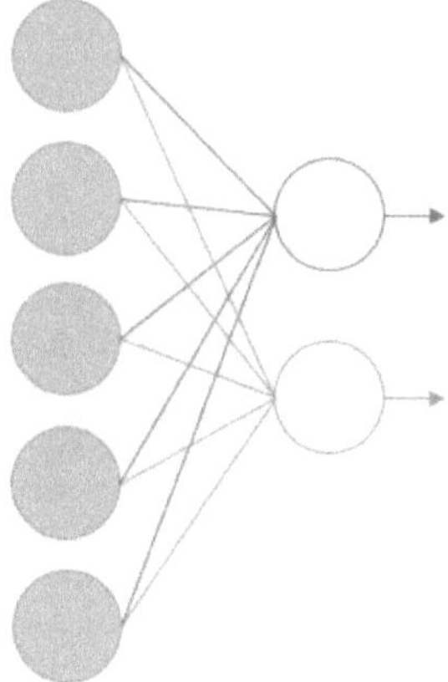

Fig. 2.2 Fully connected layer.

Fig. 2.2 demonstrates an FC layer. FC layers are widely used in classification problems and rarely used in the image restoration network of the SR model. Like the pooling layer, an FC layer can be applied to the sub-channels of the SR model. FC layer and the pooling layer can work together to form a feature fusion block. Specifically, we can use a pooling layer to compress the feature map into $1 \times 1 \times C$. The FC layer is then utilized to map each channel to a label or weight. This feature fusion block composed of a pooling layer and an FC layer is widely used in image classification and attention mechanisms.

3D Convolutional Layer 3D convolution is an extension of 2D convolution. It was initially used in behavior recognition [38] and was widely used because of its superiority in the process of consecutive images. In the 2D convolutional layer, the size of the input matrix and the filter is $M \times N$. In contrast, in the 3D convolutional layer, the input matrix and filter size can be expressed as $M \times N \times T$, where T denotes the temporal information. Fig. 2.3 compares 2D convolution operation and 3D convolution operation. Benefiting from the information on the time component, 3D convolution estimates the motion of objects between multiple frames. However, 3D convolutional layers also bring a large number of parameters to the network.

3D convolution aims to calculate low-level features in videos or 3D images. The output of the 3D convolutional layer is generally a 3D volume space. We can reduce the dimensionality of the output of the 3D convolution to a 2D feature map, which can be used as the input of 2D convolution. We can then use 2D convolution layers to obtain high-level features. 3D convolutional layers as the main structure of motion estimation are widely deployed in our model. To reduce the computational burden of 3D convolution and deploy multi-kernel

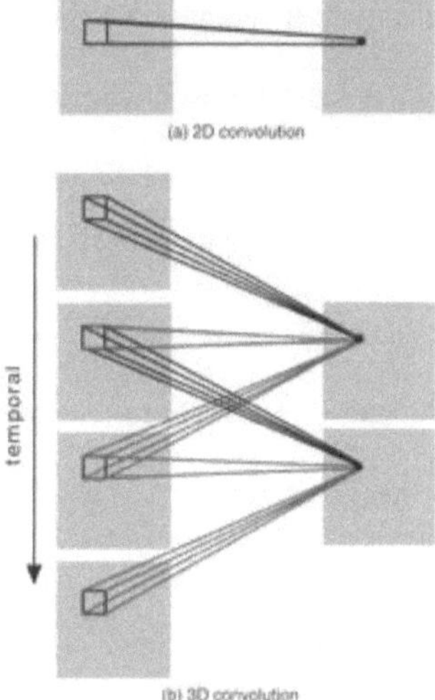

Fig. 2.3 Comparison of 2D convolution and 3D convolution.

structures, we adopt separate 3D convolution instead of regular 3D convolution layers. In

contrast to the original separate 3D convolution structure proposed by Tran et al. [39],

which first performs spatial convolution iteration and then performs temporal convolution

iteration, we alternate between the training of temporal and spatial convolutions to facilitate

multi-channel information fusion based on attention mechanism.

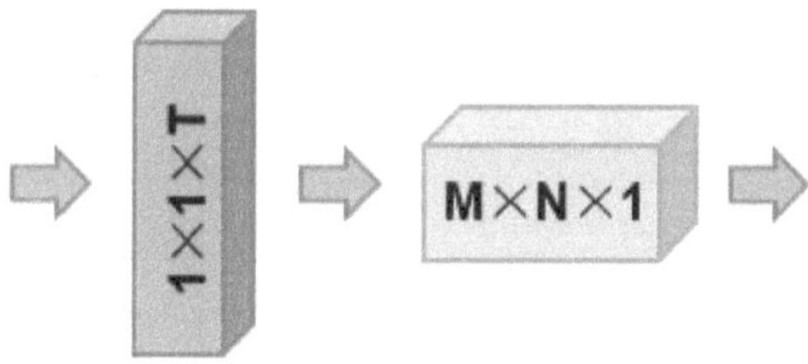

Fig. 2.4 Separate 3D convolution operation.

Fig. 2.4 illustrates the operation of separate 3D convolution. It can be seen that a separate 3D convolution decomposes a 3D convolution into a 2D spatial convolution and a 1D temporal convolution. The size of the original 3D convolution is expressed as M × N × T. The 2D spatial convolution in a separate 3D convolution can be expressed as M × N × 1, and the 1D temporal convolution can be expressed as 1 × 1 × T. Therefore, the computational parameters of separate 3D convolution can be roughly regarded as M × N + T, which significantly reduces the parameters of M × N × T for 3D convolution.

2.1.2 Activation Function

The purpose of an activation layer is to enhance the non-linear mapping capability of the network. A convolutional layer is usually accompanied by an activation layer. By introducing non-linear factors, the neural network has the ability to approximate non-linear functions. Thanks to the activation layer, as the network deepens, the network has stronger non-linear characterization capabilities, which enables deep neural networks to capitalize on their advantages. The commonly used activation functions include Sigmoid, ReLU, and Tanh.

Sigmoid The purpose of the Sigmoid function is to transform the input to values from 0 to 1, so the Sigmoid function is suitable for networks that estimate the probability. An example of the Sigmoid function is illustrated in Fig. 2.5.

The Sigmoid function is an S-shaped curve and is symmetrical at the center at a value of 0.5 [40]. The Sigmoid function can be calculated as follows:

$$\theta(x) = \frac{1}{1 + e^{-x}}.$$

(2.2)

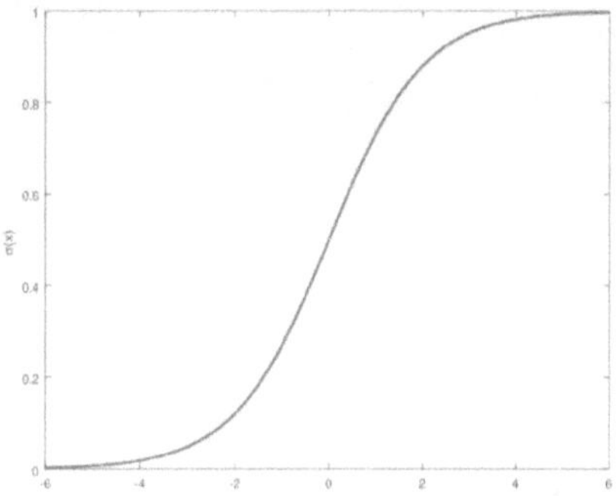

Fig. 2.5 Sigmoid function.

The Sigmoid function is not zero-centered. As the network deepens, it will change the original distribution of the data.

Tanh Unlike the Sigmoid function, the Tanh function has a value range between -1 and 1, so the output of the Tanh is zero-centered. The Tanh function can be expressed as:

$$Tanh(x) = \frac{sinh(x)}{cosh(x)}.$$ (2.3)

The derivative range of the Tanh function is between 0 and 1. Compared with the derivative range of the Sigmoid function, which is between 0 and 0.25, the gradient vanishing problem will be alleviated, but it will still exist.

Fig. 2.6 shows the Tanh function. It can be seen that both the Sigmoid function and the Tanh function will have a derivative of 0 on both sides of x, tending to infinity. Activation saturation is thus possible to happen, which may cause the gradient vanishing.

ReLU The ReLU function solves the problem of the activation saturation of the Sigmoid

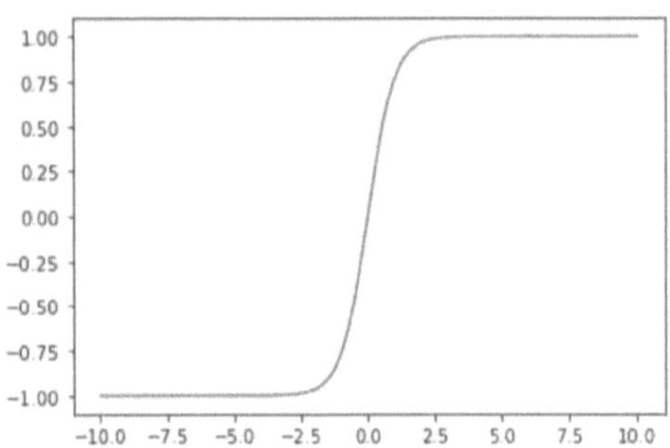

Fig. 2.6 Tanh function.

function and the Tanh function. The ReLU function is shown in Fig 2.7. It is defined as:

$$f(x) = max(0, x). \tag{2.4}$$

Fig. 2.7 demonstrates the ReLU function. ReLU has shown to be more similar to biological neurons [41]. Compared with Sigmoid and Tanh, ReLU has a smaller computational complexity. ReLU function returns the value provided as input directly when the input value is greater than zero, which avoids gradient vanishing. Moreover, [42] proves that using ReLU activation with simple formula can simplify calculations and accelerate convergence. If the input value of the ReLU function is zero or less, the output will be zero. However, this may cause some nodes to constantly output zero values. Therefore, we should avoid using a large learning rate when using ReLU as the activation function.

The improved version of ReLU optimizes the negative part of ReLU. For example, ELU [43] replaces the negative part of ReLU with: $ELU(x) = \alpha(e^x - 1)$; LReLU replaces the negative part with: $LReLU(x) = \alpha x$.

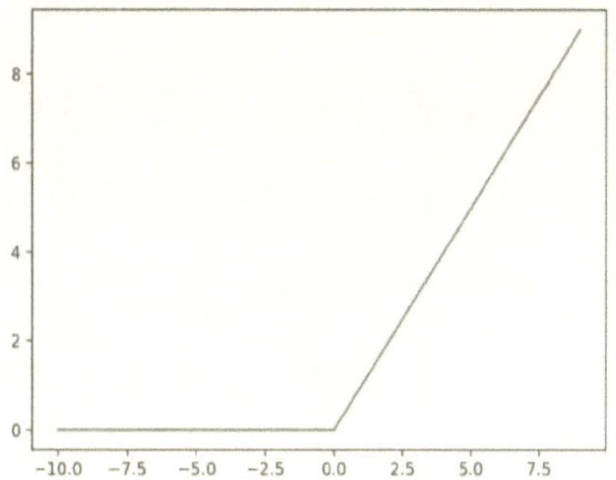
Fig. 2.7 ReLU function.

2.1.3   Loss Function

The loss function is used for model parameter estimation in machine learning. It is crucial to choose the appropriate loss function according to different machine learning tasks [34].

Mean absolute error (MAE) and mean square error (MSE) are currently commonly used loss functions in machine learning. MAE is also called L1 Loss, which is the sum of the absolute value of the difference between the target value and the predicted value. It represents the average error margin of the predicted value. MAE is defined as follows:

$$L_{MAE} = \frac{1}{N} \sum_{i=1}^{N} |y_i - F(x_i)| , \tag{2.5}$$

where N denotes the number of input data, y_i denotes the reference value, and $F(x_i)$ denotes the output value. MSE is defined as:

$$L_{MSE} = \frac{1}{N} \sum_{i=1}^{N} (y_i - F(x_i))^2. \tag{2.6}$$

The range of MAE and MSE is from 0 to ∞. MAE and MSE are primarily used for regression tasks in deep learning. MSE is more sensitive to errors than MAE. Most VSR methods use

MSE as the loss function.

Cross-entropy is generally employed as a loss function for classification tasks. In image processing, cross-entropy is used to measure the reference distribution and model prediction distribution. The formula of cross-entropy is shown below:

$$L_{CrossEntropy} = \sum_{i=1}^{N} p(x_i)log(q(x_i)),$$ (2.7)

where $p(x_i)$ denotes the reference probability distributions. $q(x_i)$ denotes the predictive probability distributions and N denotes the number of input data.

2.2   CNN Training

The purpose of a CNN is to build a mapping between input and output. CNN training is the process of adjusting various parameters in the CNN model to achieve optimal mapping. CNN training can be divided into three steps: initialization, forward propagation, and backward propagation.

Initialization The initialization of weights is essential for network training. Proper initialization parameters can speed up the convergence of the model and can help the CNN to achieve the optimal mapping. In our experiments, using different initialization methods affects the performance of the trained model.

The simple initialization method is random initialization, in which the initial weights of the model are set to random values. Random initialization causes the distribution of the network output data to change with the number of neurons. Random initialization is a commonly used initialization method for early neural networks. However, an initialization weight that is too large or too small will cause the loss gradient to be too large or too small to flow backward effectively. An initialization weight that is too large will cause the output of the layer to have a large value, which will lead to unstable training or even layer activation output explosion. Small initialization weights are suitable for small networks. For large networks, small weights lead to small gradients in backward propagation calculations, thus resulting in slow convergence and ineffective optimization.

Xavier initialization [44] is currently a commonly used method of neural network weight initialization. The Xavier algorithm automatically determines the range of initialization based on the number of input and output neurons, thereby reduces the gradient dispersion. With the Xavier algorithm, the weights in each layer are distributed with a mean value of zero and a specific variance. The variance can be defined as:

$$Var(w_i) = \frac{2}{n_{in} + n_{out}},$$

(2.8)

where n_{in} and n_{out} are the number of input channels and output channels, respectively. Therefore the weight sampling interval can be:

$$w_i \sim U\left[-\frac{\sqrt{6}}{\sqrt{n_{in} + n_{out}}}, \frac{\sqrt{6}}{\sqrt{n_{in} + n_{out}}}\right],$$

(2.9)

where $U[-x, x]$ refers to the uniform distribution in the interval $(-x, x)$.

Similar to the Xavier method, He initialization [45] aims to maintain the variance of the state value during forward propagation and maintain the variance of the gradient during backward propagation. In contrast to Xavier initialization, He initialization takes into account the influence of ReLU activation on the output data distribution. When the activation function is ReLU, the variance is:

$$Var(w_i) = \frac{2}{n_{in}},$$ (2.10)

where n_{in} is the number of input channels.

Batch normalization (BN) [46] aims to normalize the activation in intermediate layers of deep neural networks [47]. BN sets the distribution of the input value of each intermediate layer to a standard normal distribution with a mean of 0 and a variance of 1. Therefore, using the BN layer in the intermediate layers of the network can reduce the network's dependence on initialization. Specifically, the BN of each dimension can be expressed as :

$$\widehat{x}^{(k)} = \frac{x^{(k)} - E[x^{(k)}]}{\sqrt{Var[x^{(k)}]}},$$ (2.11)

where k denotes the kth dimension. E and Var denote expectation and variance, respectively, and they are computed through the training dataset [46]. BN not only makes CNN more stable but also can effectively accelerate training [48]. In addition, BN allows higher learning rates. These advantages make BN widely used in deep networks.

Forward Propagation The forward and backward propagation in deep learning simulate the forward and reverse signal conduction of human brain neurons [49]. Most of the existing neural networks train models through the loop of forward propagation and backward propagation. Forward propagation is the process of calculating the output from the input data and the neural network. The forward propagation can be expressed as:

$$F(x) = f_n(f_{n-1}...f_1(x)),$$ (2.12)

where x denotes the input of the network, and F(x) denotes the output of the final layer.

Backward Propagation Backward propagation (BP) allows information to flow backward through the network from the loss value to calculate the gradient efficiently. BP uses the response obtained by forward propagation to calculate the response error of the hidden layer and the output layer. If the nested function is $F(x) = f_n(f_{n-1}...f_1(x))$, the derivative of the input x can be expressed as:

$$\frac{dF}{dx} = \frac{df_n}{df_{n-1}} \cdot \frac{df_{n-1}}{df_{n-2}} \cdot ... \cdot \frac{df_1}{x}.$$

(2.13)

The loss value L can be calculated according to the chain rule. For example, for node i in layer k, the partial derivative of the loss with respect to weight parameter w_i^k can be calculated by:

$$\frac{\partial L}{\partial w_i^k} = \frac{\partial L}{\partial \alpha_i^k} \cdot \frac{\partial \alpha_i^k}{\partial \beta_i^k} \cdot \frac{\partial \beta_i^k}{\partial w_i^k},$$

(2.14)

where α_i^k represents the output of the activation layer of node i in layer k. β_i^k is the output of the convolutional layer of node i in layer k. It is proven by [50] that backward propagation enables networks to have a higher learning rate. Thus backward propagation becomes a crucial part of a CNN.

Optimization The goal of optimization is to minimize the loss function. The training of the parameters is expected to reach the global minimum of the loss function. However, it is also possible to reach the local minimum, which results in the parameters being unable to reach the optimal. Gradient descent is the core of neural network optimization, and optimization methods are strategies for different networks based on gradient descent. The standard gradient descent can be expressed as:

$$\theta_{t+1} = \theta_t - \eta \cdot \nabla J(\theta_t),$$

(2.15)

where θ denote the parameters, η denotes the learning rate, and $\nabla J(\theta_t)$ is the gradient loss with respect to θ_t. The standard gradient descent should be calculated once on the entire data set for each iteration, which leads to slow gradient descent and requires large memory.

Stochastic gradient descent (SGD) updates the parameters of each training statistic [51]. SGD can achieve accelerated convergence speed and does not require large memory. SGD is defined as:

$$\theta_{t+1} = \theta_t - \eta \cdot \nabla J(\theta_t; x_i, y_i), \tag{2.16}$$

where x_i and y_i denote the training input data and the corresponding output data. However, updating the weights for each data increases the possibility of reaching the local optimum.

Gradient descent with momentum is introduced to ensure that the networks converge stably [52]. The momentum algorithm uses the concept of momentum in physics. It simulates the inertia of an object, meaning that it retains the previously updated direction to a certain extent when updating. Therefore gradient descent with momentum optimizes training in related directions and weakens oscillations in unrelated directions by updating a vector γ. Gradient descent with momentum can be defined as:

$$\triangle \theta_{t+1} = \gamma \cdot \triangle \theta_t + \eta \cdot \nabla J(\theta_t). \tag{2.17}$$

The Nesterov algorithm provides the momentum algorithm with the predictive ability [52]. Specifically, it calculates $\theta_t - \gamma V_t$ to obtain the approximate parameter value of the next position. The Nesterov algorithm can be defined as:

$$\triangle \theta_{t+1} = \gamma \cdot \triangle \theta_t + \eta \cdot \nabla J(\theta_t - \gamma V_t). \tag{2.18}$$

The AdaGrad algorithm [53] aims to automatically adjust the learning rate. It is able to make extensive updates to sparse data and small updates to frequent data. The parameter-updating function of the AdaGrad algorithm is:

$$\theta_{t+1} = \theta_t - \frac{\eta_0}{\sqrt{s_t + \epsilon}} \cdot g_t, \tag{2.19}$$

where η_0 is the initial learning rate, ϵ is a constant to avoid zero denominators. g_t is the gradient estimation of θ at time step t. And s_t is the sum of squared gradient over the training process, which is calculated as:

$$s_t = \sum_{\tau=1}^{t} (g_\tau)^2. \tag{2.20}$$

However, the Adagrad algorithm always reduces the learning rate, resulting in slow training in the later stage of training and an early end to the training.

The Adadelta algorithm solves the problem of the continuous decay of the AdaGrad algorithm learning rate. Instead of calculating the summation of all past squared gradients, the Adadelta algorithm recursively calculates the sum of all the previous squared gradients. The sum of the previous gradient is calculated as:

$$s_{t+1} = \alpha \cdot s_t + (1 - \alpha) \cdot g_t^2, \tag{2.21}$$

where α is the momentum that determines the percentage of the previous squared gradients to be calculated. The Adadelta algorithm can achieve a high training speed in the later stage of training.

The Adam [54] optimization algorithm is a stochastic approximation method based on adaptive estimation of first-order and second-order moments. Similar to Adadelta, Adam retains the exponential decay average of past squared gradients. It also keeps the exponential decay average of the past gradient similar to the gradient descent with momentum. The update of Adam parameters can be defined as:

$$\theta_{t+1} = \theta_t - \frac{\eta}{\sqrt{\hat{v}_t} + \epsilon} \hat{m}_t,$$

(2.22)

where $\hat{m}_t$ and $\hat{v}_t$ are defined as:

$$\hat{m}_t = \frac{m_t}{1 - \beta_1^t},$$

(2.23)

$$\hat{v}_t = \frac{v_t}{1 - \beta_2^t},$$

(2.24)

where m_t and v_t are the estimated value of the first moment (the mean) and the second moment (the variance) of the gradients, respectively. β_1^t and β_2^t denote the decay rate of the first moment and the second moment, respectively. The default value of β_1^t and β_2^t are set to 0.9 and 0.999 respectively. Adam's default parameters are suitable for most networks, making Adam widely used in various types of deep learning tasks.

2.3 CNN Related Networks

In this section, we first list the existing multi-receptive field CNNs. We then introduce the residual network [3] as an essential breakthrough in deep neural networks. Next, we introduce the dilated convolution, which is used to build multi-receptive field structures in our work. After that, we present deformable convolution, which is utilized for multi-frames alignment. Finally, we review the Spatial Transformation Network that enables VSR models to effectively integrate time and space information.

2.3.1   Multi-Kernel CNN

The receptive fields of standard convolutions share the same size in each layer. However, the input image contains different objects of interest, which results in the receptive field of the same size being insufficient to adjust according to various inputs. Some research has started to focus on deploying different convolution kernels in the same layer.

Inception V1 [1] splices convolution kernels of different sizes in the same layer to provide different receptive fields. The use of convolution kernels of different sizes improves the performance of the network of limited computing resources. Inception V1 won the ILSVRC14 of the year with its two advantages. Firstly, the structure of multiple convolution kernels is suitable for recognition tasks. In the recognition task, there are differences in the position of the information. Images with more global information distribution prefer larger convolution kernels, and images with more local information distribution prefer smaller convolution kernels. Secondly, before the birth of the residual structure, a deep network is difficult to be applied because of the problem of gradient explosion and vanishing. Therefore, the wider structure of Inception V1 can achieve good performance while avoiding the use of deep networks. Fig. 2.8 shows the naïve version of Inception V1.

Subsequently, Google proposed a subsequent version of Inception. Inception V2 proposes batch normalization, which can significantly accelerate the convergence speed of large convolutional networks, and has become a milestone for CNNs. Inception V3 splits a larger 2D convolution into two smaller 1D convolutions. Not only does this reduce the computational burden, speed up the calculation, and reduce over-fitting, it also increases the non-linearity of

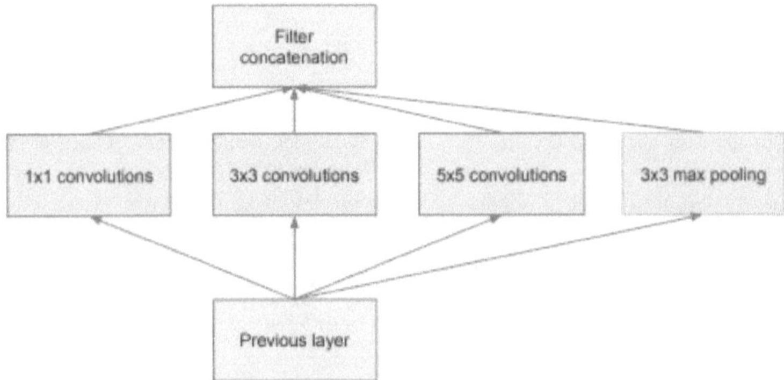

Fig. 2.8 Architecture of Inception-V1 (adopted from the original paper [1]).

the model. In Inception V4, the multi-convolution kernel structure and the residual structure are combined to achieve a deeper network and better performance in visual recognition.

Inception takes the equivalent combination of each convolution kernel as the input of the next layer. Unlike Inception, Selected Kernel Network (SKNet) [2] uses the attention mechanism to fuse each convolution kernel according to the different inputs. SKNet is inspired by cortical neurons that can dynamically adjust their own receptive field according to different stimuli. Fig. 2.9 shows the structure of the SKNet.

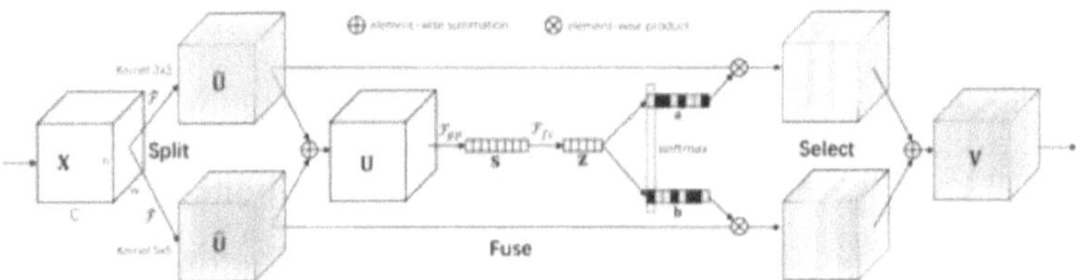

Fig. 2.9 Architecture of SKNet (adopted from the original paper [2]).

SKNet applies the channel attention mechanism proposed in SENet [7] to the weight distribution of different convolution kernels. In addition, SKNet introduced dilated convolution to form large convolution kernels.

Inspired by these works, we propose multi-kernel structures suitable for VSR tasks.

2.3.2   ResNet

A deep network can achieve a larger receptive field and thus obtain better performance than a shallow network. However, deep networks are difficult to train. Even if a large amount of data can solve the problem of network overfitting, batch normalization can also prevent the gradient from vanishing or exploding. The performance of a deep network is thus not necessarily better than a shallower network.

The residual network (ResNet) [3] proposed by He, K et al. has solved this problem. The authors indicated that when the depth of the network increases after accuracy saturated, the network will degrade. When the network is degraded, the shallow network can achieve better training results than the deep network. To address this problem, the authors proposed a skip connection (residual) structure to transfer shallow features to deep layers. Fig. 2.10 shows the structure of a residual block.

The residual block avoids degradation, gradient vanishing or exploding, and overfitting through shortcut connections. Moreover, the residual block can be easily deployed in various networks. Therefore, the residual structure is widely used in various deep learning fields. We extensively use the residual structure in our work to build a deep feature extraction network.

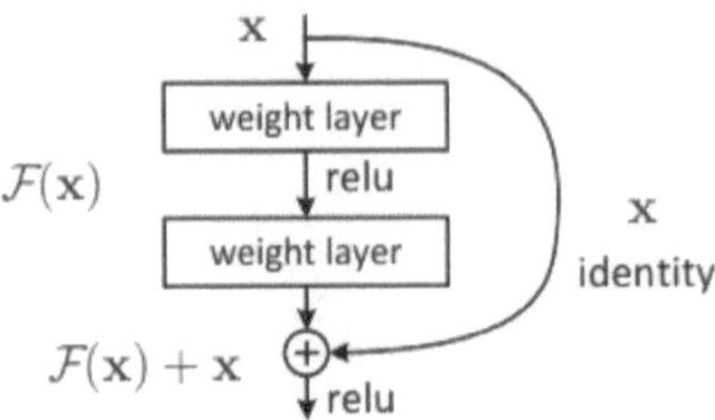

Fig. 2.10 Structure of a residual block (adopted from the original paper [3]).

2.3.3 Dilated Convolution

Dilated convolution [4] was first proposed in the field of context segmentation. Standard context segmentation methods utilize pooling layers to obtain greater receptive field information. However, the pooling layer reduces the size of the feature map, inevitably resulting in information loss [55]. The purpose of dilated convolution is to enlarge the receptive field without applying the pooling layer. Specifically, dilated convolution adds holes to each primary convolution. The standard convolution operation can be expressed as:

$$(F * k)(p) = \sum_{s+t=p} F(s)\,k(t),\qquad(2.25)$$

where s refers to stride, and p refers to the receptive field. The size of kernel k is r, and $t \in [-r, r]$. The dilated convolution operation with holes added can be expressed as:

$$(F * \mu k)(p) = \sum_{s+\mu t=p} F(s)\,k(t),\qquad(2.26)$$

where μ denotes the dilation rate. Fig. 2.11 intuitively shows the sampling points of dilated convolution. Fig. 2.11 (a) is produced from F_0 with dilated convolution when the dilation rate equals 1. The receptive field is 3×3, equivalent to a 3×3 standard convolution. Fig.

2.11 (b) is produced with dilated convolution with dilation rate 2, the receptive field is 7 × 7. Moreover, for Fig. 2.11 (c), the dilation rate is 3, and the receptive field is 15 × 15. Thus, dilated convolution expands the receptive field while avoiding a heavy computational burden.

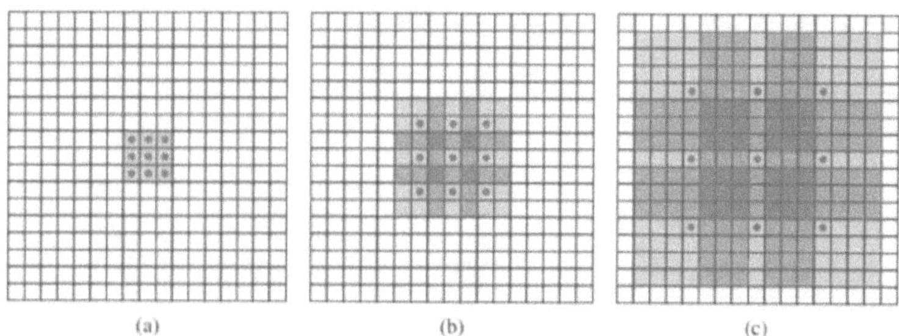

Fig. 2.11 Sampling points of dilated convolution (adopted from the original paper [4]).

In VSR tasks, the use of a pooling layer has a worse impact than context segmentation [56]. Therefore, the dilated convolution is more suitable for VSR to adjust the receptive field. In our work, we embed the dilated convolution into the multi-kernel structure to replace the large convolution kernel.

2.3.4 Deformable Convolution

CNN has limitations in modeling geometric transformations. Firstly, a standard CNN uses a fixed geometric structure to construct the model, resulting in limited geometric transformation capabilities. Secondly, in the same layer of the CNN, objects of different sizes have the same receptive field. Lastly, a pooling layer reduces the spatial resolution, and an

RoI (region of interest) pooling layer divides the RoI into fixed spatial bins, further losing resolution features.

Deformable convolution networks [5] propose two modules to solve these problems: deformable convolution and deformable RoI pooling. In this **book**, we only discuss deformable convolution that can be applied to VSR. In deformable convolution, each convolution operation has an additional offset. Specifically, the output feature map F_O of the standard convolution is:

$$F_O(p_0) = \sum_{i=1}^{K} w_i\,(p_i) \cdot F_I\,(p_0 + p_i)\,, \tag{2.27}$$

and the output feature map F_D of the deformable convolution operation can be expressed as:

$$F_D(p_0) = \sum_{i=1}^{K} w_i\,(p_i) \cdot F_I(p_0 + p_i + \Delta p_i), \tag{2.28}$$

where the offset Δp_i is obtained by applying convolutional layers over the input feature map F_I. The offset map is calculated by a parallel standard convolution unit and it can be used for end-to-end learning. Fig. 2.12 shows the operation process of deformable convolution.

Deformable convolution is able to adapt to object motion and geometric transformation. Furthermore, deformable convolution is an operation on the 2D domain, which can be extended to a 3D CNN. Thus, some VSR methods apply deformable convolution for multiple consecutive frames alignment [20, 57, 58]. The offset of each pixel between the neighboring frame and the current frame is learned through standard convolution to form an offset field. Each pixel in the 3D feature map is then rearranged by deformable convolution to align the neighboring frames.

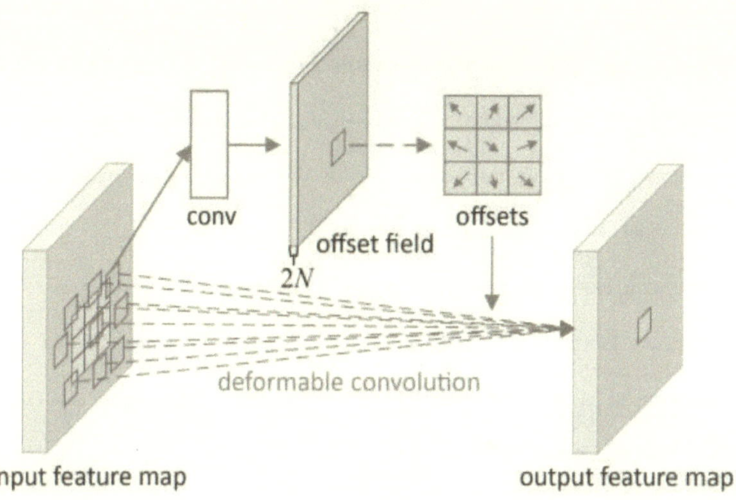

Fig. 2.12 Operation of deformable convolution (adopted from the original paper [5]).

The existing VSR methods deploy standard convolution to obtain the offset for each pixel between multiple frames, but it is difficult for standard convolution to estimate high-speed objects. To solve this problem, we propose a multi-kernel offset estimation module to increase the spatial receptive field of deformable convolution. Moreover, we further combine multi-kernel deformable convolution with separate 3D convolution to improve the quality of the alignment.

2.3.5 Spatial Transformation Network

Due to the importance of the spatial transformation network (STN) [6] to VSR, we will review the STN. The image processing model should have spatial invariance to deal with the spatial transformation of the image. However, because the space support for max-pooling is usually small, the CNN has limited space restrictions and is difficult to adapt to large spatial transformations. Therefore, STN aims to align the image as the preprocessing of the network's input. This process can be embedded in the network to construct a weight-shared end-to-end model.

Fig. 2.13 illustrates the architecture of the spatial transformer module, which includes three parts: the localization net, the grid generator, and the sampler. The purpose of the localization net is to map the coordinate relationship between input and output, and to produce the transformation parameters θ. The grid generator calculates the coordinate points of the output feature map according to the coordinates of the input and transformation parameters θ. The sampler uses interpolation to restore the resolution of the image.

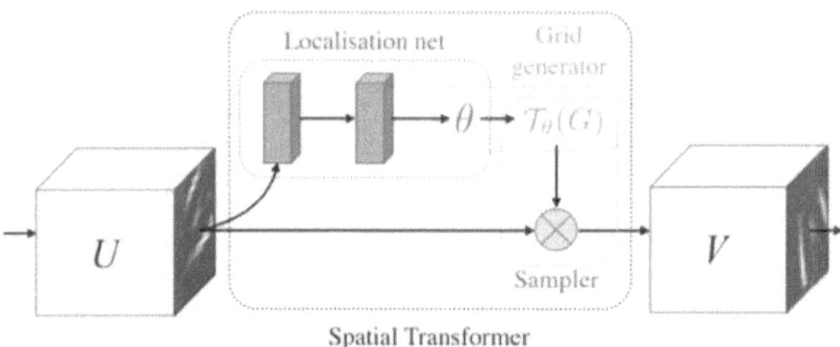

Fig. 2.13 Spatial transformer (adopted from the original paper [6]).

The spatial transformer applied to the VSR is slightly different from the original STN. In VSR, the role of the spatial transformer is to align multiple frames of images. Generally, the VSR model takes the middle frame of the input image as the current frame, and the goal of the spatial transformer is to align the neighboring frames to the current frame. The localization net is hence replaced by a motion estimation network, such as the optical flow algorithm and the 3D convolution network.

2.4 Attention Mechanism

The attention mechanism has been widely used in various fields of deep learning in recent years, such as image processing, machine translation, and natural language processing [20, 59, 32]. The attention mechanism employed in deep learning is inspired by the human visual attention mechanism, which quickly scans global images to obtain the target area. In computer vision related neural networks, the attention mechanism is primarily used for weight distribution. The sum of weights is calculated by the non-local operation of time-space locations to capture non-local dependencies.

The attention mechanism is deployed in various parts of the neural network and form a variety of attention mechanisms, including channel attention, temporal-spatial attention, pixel attention, and multi-level attention. Among these, we utilize channel attention to assign weights to each channel with different receptive fields and temporal-spatial attention to align multiple consecutive frames.

2.4.1 Channel Attention Mechanism

In a neural network, different convolution kernels in each layer of convolution can be defined as different channels. In a multi-channel network, with the help of channel attention, the learnable weight can be adjusted according to the input feature map. Squeeze-and-excitation network (SENet) [7] proposes an embedded channel attention model that can be described as two steps: squeeze and excitation. SENet first performs the squeeze operation on the feature map obtained by convolution to obtain the channel-level global features. They then perform the excitation operation on the global features to obtain the weights of different channels. Finally, the original input feature map is multiplied by the channel weights to produce the output feature map.

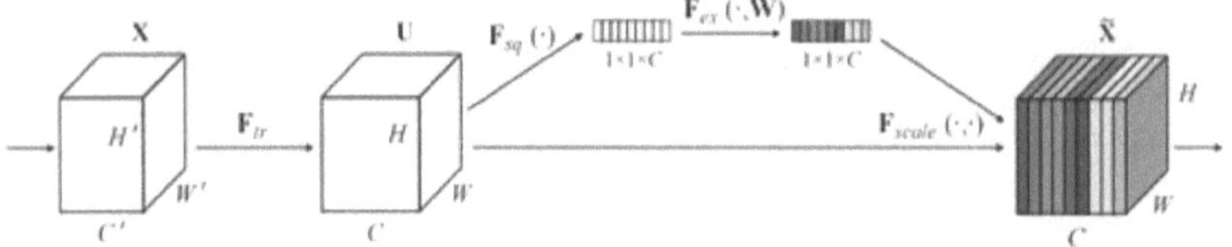

Fig. 2.14 Architecture of SENet (adopted from the original paper [7]).

Fig. 2.14 shows the two-step operation of the SENet for generating a channel attention module. This module can be combined with the residual structure to build a deep channel attention network. Fig. 2.15 shows the structure of the combination of the SENet and the residual network.

Unlike SENet, which uses average pooling to obtain global information, the convolutional block attention Module (CBAM) [59] combines maximum pooling and average pooling to improve the performance of channel attention. In the VSR study, the method proposed by CBAM that combines the two pooling operations has been shown to achieve better channel attention results.

In our work, convolution kernels of different sizes are regarded as different channels, and

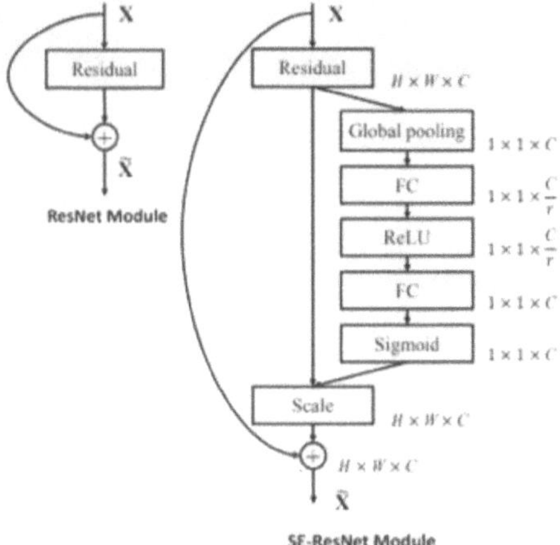

Fig. 2.15 Combination of SENet and residual network (adopted from the original paper [7]).

the channel attention mechanism is deployed in each channel. We use the sum of average pooling and maximum pooling to achieve better performance. In addition, we extend the channel attention to separate 3D convolutions to assign weights to different convolution kernels of the time-space domain.

2.4.2 Spatial and Temporal Attention Mechanisms

Temporal attention mechanisms and spatial attention mechanisms are explained differently in different deep-learning projects. In VSR, the spatial attention mechanism aims to align multiple consecutive frames through spatial transformation, and the temporal attention mechanism aims to assign weights to each frame.

In Sec. 2.3.5, the spatial transformation network that we introduced can be regarded as the pioneering work of spatial attention, because the trained spatial transformer has the ability to find the area of interest in the picture. In the VSR task, spatial attention is mainly used to compare the relationship between each neighboring frame and the current frame so as to assign different weights to each frame. Previous VSR studies have proven that time and space attention mechanisms can improve the performance of the VSR model [59].

Chapter 3

Literature Review

Deep-learning-based SR methods have attracted considerable attention in recent years due to their superior performance. The development of deep CNNs in all aspects has caused learning-based SR to have the potential for continuous improvement. In this chapter, we will review deep-learning-based SISR methods and deep-learning-based VSR methods.

3.1 Single Image Super-Resolution

As the pioneering work of learning-based SR, SRCNN [8] deploys three convolutional layers for feature extraction, non-linear mapping, and feature reconstruction. Although a simple network structure is used in SRCNN, the experimental results greatly surpass the traditional methods. Fig. 3.1 demonstrates the structure of these three convolutional layers. In addition to being the first to combine CNN with SR, the experimental procedures of SRCNN have also become a template for subsequent SR research. For data preprocessing, LR images

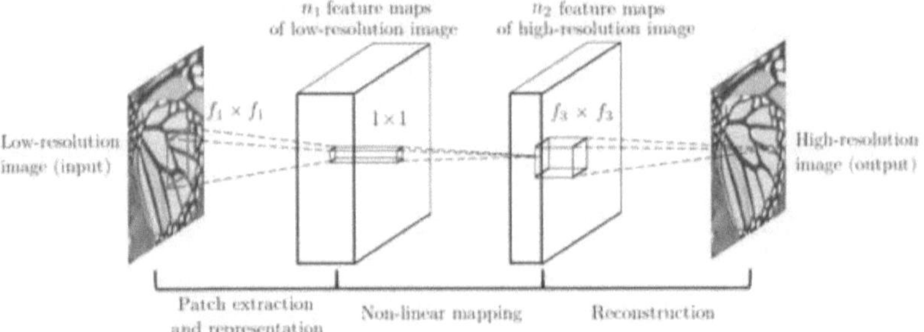

Fig. 3.1 Architecture of SRCNN (adopted from the original paper [8]).

are first subjected to bilinear interpolation to the same size as HR images. The non-linear mapping is then fit through a three-layer convolutional network, and finally, the SR image is generated as the output. Since PSNR has always been used as an SR evaluation index, SRCNN uses MSE as the loss function of the network.

FSRCNN [18] is an improved version of SRCNN. FSRCNN uses a deconvolution layer to enlarge the size of LR images instead of using bilinear interpolation. Therefore, the original LR image is directly fed into the network, reducing the computational cost. FSRCNN adopts smaller convolution kernels, so FSRCNN is faster than SRCNN. Moreover, FSRCNN uses more mapping layers to replace the three-layer structure of SRCNN. Fig. 3.2 compares the structure between SRCNN and FSRCNN.

Since SRCNN adopts the interpolated LR picture as the network's input, more calculations are required to convolve these large input images, resulting in a heavy computational burden. ESPCN [10] proposes sub-pixel convolutional layers for extracting features directly on LR images to reduce the computational cost. ESPCN first uses two convolutional layers to obtain the multi-channel LR feature maps and then rearranges the multiple LR feature

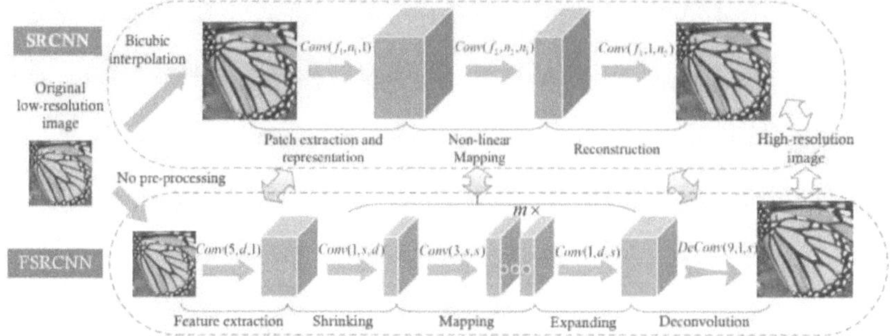

Fig. 3.2 Comparison of SRCNN and FSRCNN (adopted from the original paper [9]).

maps' pixels to produce the sub-pixels of the SR output. ESPCN is a real-time SISR network and one of the fastest SISR methods. Fig. 3.3 demonstrates the process of sub-pixel convolutional neural network.

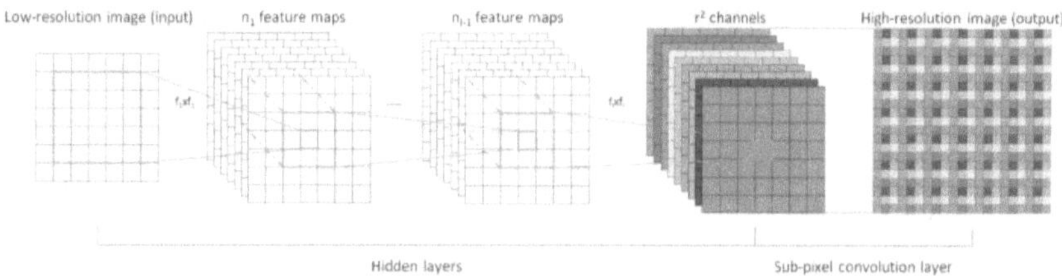

Fig. 3.3 The process of sub-pixel convolutional neural network (adopted from the original paper [10]).

Before ResNet was employed, deep networks were difficult to train, which limited the performance of learning-based SR networks. The residual blocks are suitable to be deployed in the SR models, because LR images are similar to HR images in low-frequency parts. Learning high-frequency residuals can make the networks converge faster. With the help of VGG-net [60], VDSR [11] adopts residual structures to reach 20-weight layers to increase the receptive field. VDSR is the first to apply zero-padding to the SR network so that the size of the output image is the same as the input. Moreover, VDSR also proves that zero-padding can improve the quality of the restoration of boundary pixels. Fig. 3.4 shows the residual structure of VDSR. The experimental result of VDSR outperforms SRCNN by

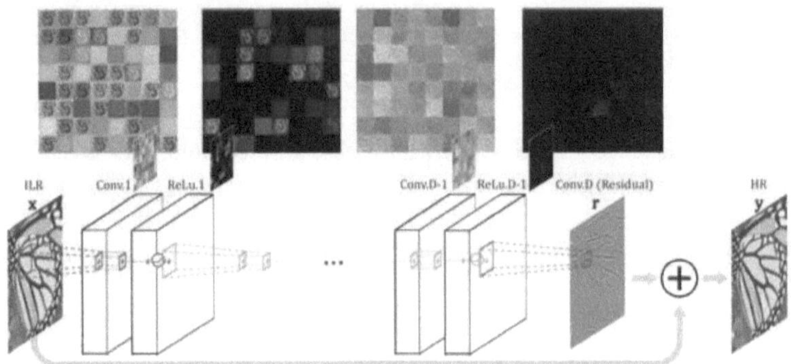

Fig. 3.4 Architecture of VDSR (adopted from the original paper [11]).

a large margin, which proves that the deep network can achieve better performance in SR. In addition, VDSR mixes images of different scales for training to solve the SR problem of different scales.

RED [12] first applies the encoding-decoding structure to SR. As shown in Fig. 3.5, the RED network is symmetrical, and each convolutional layer has a corresponding deconvolutional layer [61]. Furthermore, RED connects each set of corresponding convolutional layer and deconvolutional layer with a skip connection. The feature maps' sizes are shrunk by the convolutional layers without padding, and the deconvolutional layers then recover the feature maps to the same size as input. The symmetrical and skip-connected structure allows the back-propagating information to be directly passed to the bottom layers, making more detail delivered and making the network easier training.

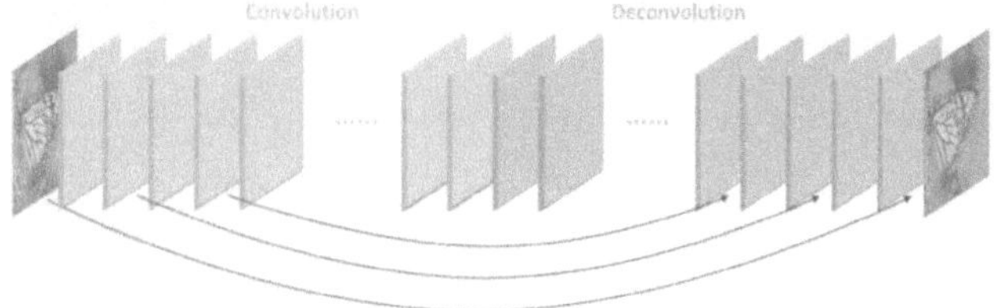

Fig. 3.5 Architecture of RED (adopted from the original paper [12]).

The attention mechanism has achieved great success in the field of natural language processing. In recent years, some studies have introduced the attention mechanism into SR. RCAN (residual channel attention network) [13] is one of the first successful architectures solving image SR tasks using a channel attention model. They also combine channel attention with a residual network to form a residual channel attention block, which has achieved remarkable success on Set5 and Set14, proving that for SR images, the channel attention model can be well integrated into the residual network. Furthermore, RCAN proposed the residual in residual structure (RIR) to build a very deep trainable network, making the network learning more effective. Fig. 3.6 shows the architecture of RCAN.

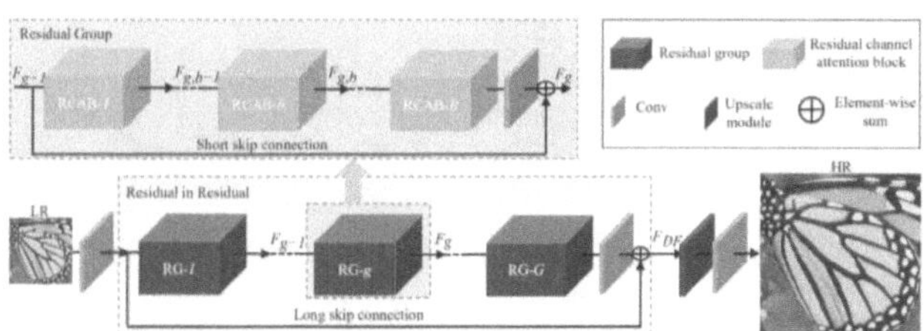

Fig. 3.6 Architecture of RCAN (adopted from the original paper [13]).

SRGAN [14] is the first method to apply the generative adversarial network (GAN) to

SISR and achieve good performance. SR methods generally use MSE as the loss function. The author of SRGAN believes that using MSE as the loss function can achieve a high PSNR, but it will cause the HR images to lose high-frequency information, leading to the HR images lacking visual realism. Therefore, SRGAN aims to restore visually real images by using GAN. GAN is composed of two sub-networks: a generation network and a discriminator network. The generation network can restore images with high PSNR testing results, and the purpose of the discriminator network is to optimize the HR image to make it visually realistic. MSE is used as the loss function to train the generation network, and the perceptual loss [62] is adopted to train the discriminator network. Fig. 3.7 illustrates the architecture of the generator and discriminator networks.

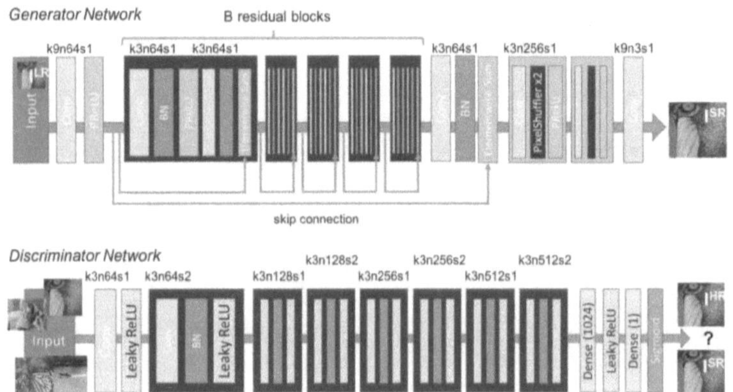

Fig. 3.7 Architecture of Generator and Discriminator Network (adopted from the original paper [14]).

3.2 Video Super-Resolution

The early CNN-based SR methods were able to restore HR videos frame by frame, but could not effectively take advantage of continuous frames to restore a higher quality video than SISR.

VSRnet [15] is considered as the first method to utilize multi-frame as the input to build a deep CNN. VSRnet divides the motion compensation part and the SR distribution into two independent networks. First, the optical flow estimation [63] is utilized to learn the motion of multi-frame objects. The multi-frame images that motion compensated are then used as the input for the super-resolution network. In addition, VSRnet first trained the SISR model as a pre-training step, before transferring the weight of the pre-training model to the VSR model. VSRnet can be considered as a method to extend the SISR method to VSR. The architecture of VSRnet is shown in Fig. 3.8. Due to the effective use of multi-frame information, VSRnet indicates a significant performance improvement compared to the SISR methods. However, due to its two-part independent structure, the advantages of CNN in VSR are unable to reach their full potential.

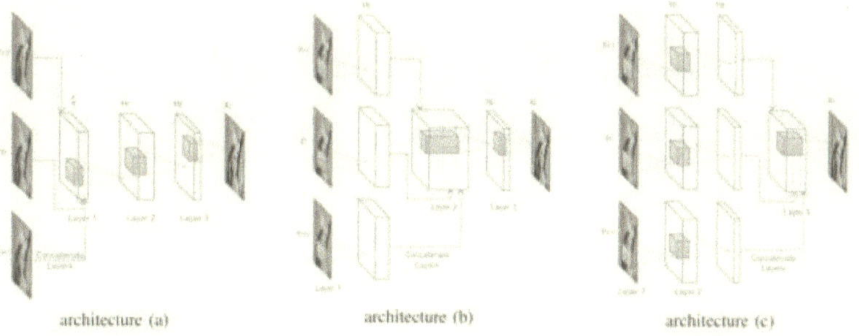

Fig. 3.8 Architecture of VSRnet (adopted from the original paper [15]).

VESPCN [16] combines SR reconstruction and multiple frames alignment to build an end-to-end joint training model. Like VSRnet, VESPCN uses optical flow for motion estimation and compensation. The architecture of VESPCN is shown in Fig. 3.9. The innovation of VESPCN is using the spatial transformation network [6] to combine the timing information of multiple frame input with the spatial information of the VSR network. This method is a great inspiration for subsequent VSR research.

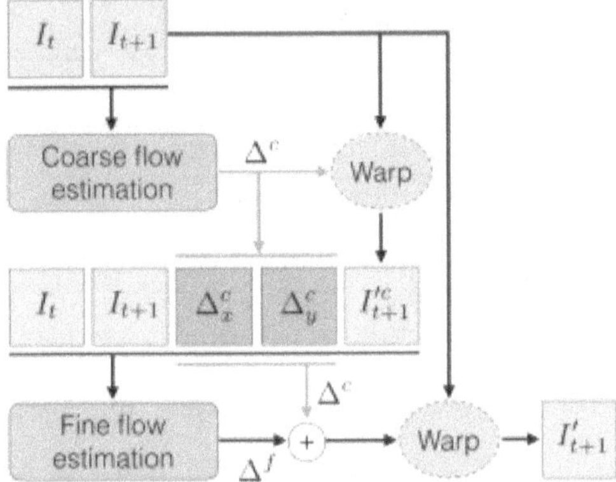

Fig. 3.9 Spatial transformer motion compensation (adopted from the original paper [16]).

SPMC [17] introduces sub-pixel motion compensation layers to process motion compensation resolution enhancement. Fig. 3.10 illustrates the sub-pixel motion compensation layer. SPMC consists of three main networks: motion estimation network, motion compensation network, and detail fusion network. The motion estimation network structure is similar to VESPCN, and optical flow is adopted to align multiple frames. The estimated flow is then utilized to calculate the transformed coordinates. The sampler combines the gridded LR frame with transformed coordinates to generate the aligned feature maps. The detail fusion network is designed as an encoding-decoding form. The convolutional layers reduce the resolution, processes the associated information within the frame, and deconvolution layers restore the size of the feature maps.

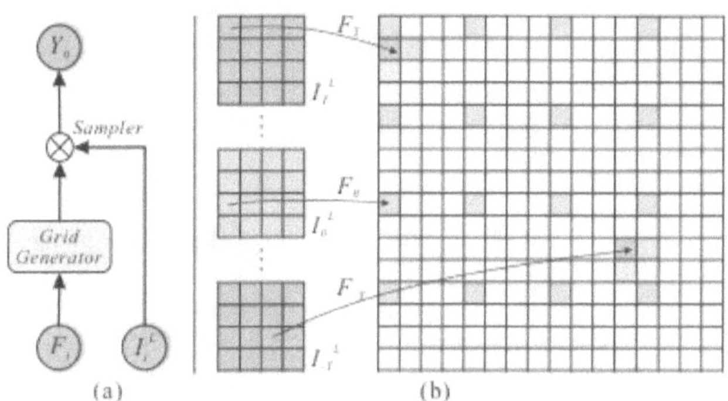

Fig. 3.10 Subpixel motion compensation layer (adopted from the original paper [17]).

Before FRVSR [18], VSR methods process a batch of LR frames to generate the SR output. However, this fashion will cause each frame to be processed multiple times, increasing the computational burden. FRVSR proposes a frame-recurrent structure, iterative taking the previously estimated HR frame as input for subsequent frames. Fig. 3.11 demonstrates the framework of FRVSR. FRVSR first uses FNet to compute optical flow, and the HR estimation of previous frames is warped onto the current frame. FRVSR then uses the space-to-depth transformation to map the warped output. Finally, the SRNet is trained for SR reconstruction. For the frame-recurrent method, each input frame is processed once, reducing the computational cost. Furthermore, the frame-recurrent method can assimilate a large number of frames without a heavy burden. Inspired by FRVSR, some VSR studies began to use the frame-recurrent structure to extract multi-frame information.

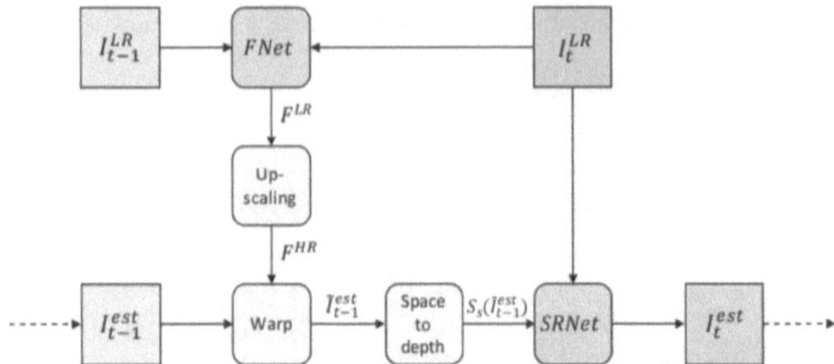

Fig. 3.11 Framework of FRVSR (adopted from the original paper [18]).

Duf [19] proposed dynamic upsampling to avoid explicit motion compensation. Dynamic upsampling filters a re g enerated b ased o n t he s patio-temporal d omain o f e ach p ixel i n LR frames. Fig. 3.12 shows the process of dynamic upsampling. Seven consecutive frames are fed into the dynamic generate network. The dynamic generate network then generates r^2HW filters. Each pixel value in the output is obtained by convolution of 16 filters (upsampling scale of 4). VSR-DUF makes the upsampling a learnable process so that multi-frame information can be effectively fused.

EDVR [20] is considered in our **book** as the previous state-of-the-art VSR method. Fig. 3.13 demonstrates the structure of EDVR. EDVR contains four modules: the pre deblur module, the PCD align module, the TSA fusion module, and the reconstruction module. The pre-deblur module builds a pyramid structure similar to the encoder-decoder network to deblur the input image. The PCD align module uses deformable convolution to complete image alignment, avoiding optical flow estimation. The TSA fusion module introduces the Attention mechanism to deploy different feature maps with different weights in the spatial

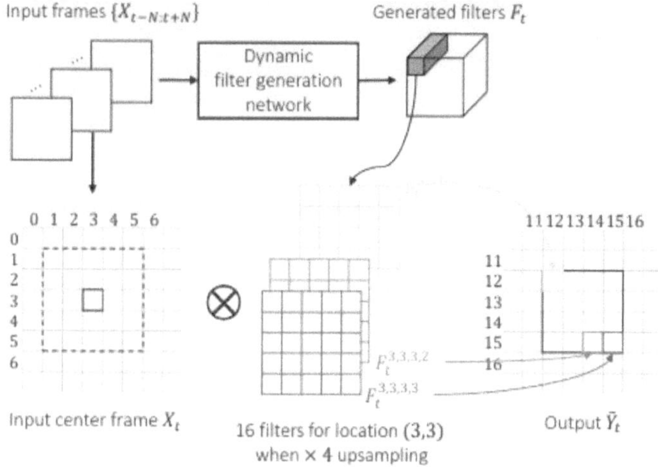

Fig. 3.12 Dynamic upsampling (adopted from the original paper [19]).

and temporal dimensions. With its excellent performance, EDVR won the first place with a considerable advantage in the CVPR NTIRE 2019 image/video enhancement competition.

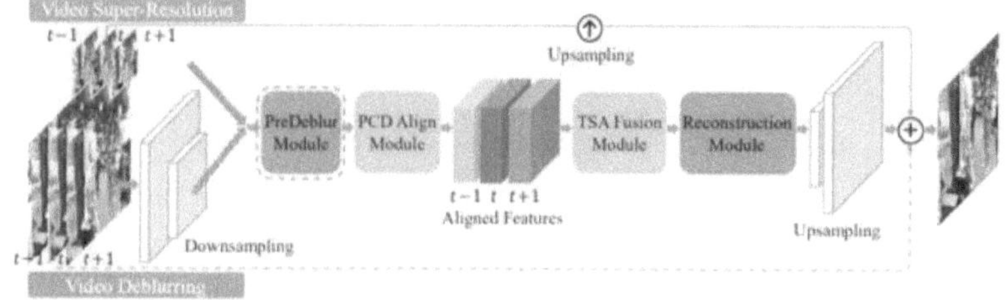

Fig. 3.13 Framework of EDVR (adopted from the original paper [20]).

3.3 Summary

In this chapter, we have reviewed the most influential deep-learning-based SISR and VSR methods, which have achieved considerably good performance. However, there is still room for VSR to improve the super-resolution quality. Our book aims to achieve a higher super-resolution quality than the previous methods. Moreover, visual cortex cells adjust the size of their receptive fields by the stimulus. None of the previous VSR methods considered to adjust the receptive field in the same layer. We will try to imitate the adjustable receptive field of biological vision to achieve more visually realistic HR videos.

Chapter 4

Video Super-Resolution via Multi-Kernel Deformable 3D Convolution

VSR tasks can be divided into two dimensions: the time dimension and the space dimension. Therefore, we can directly consider that there are three main difficulties in VSR:

1. Motion estimation in the time dimension: In VSR, we use multiple continuous images to enhance the quality of one of the images. In this operation, multiple frames need to be aligned to facilitate model training. However, different objects in the video have various motions, and determining how to estimate these motions has become the main difficulty in the time dimension.

2. Feature extraction in the space dimension: After motion estimation, we flatten the 3D feature map into a 2D feature map. Then, we use deep 2D convolutional networks to obtain features in the space dimension. However, the same object in different frames has different scales. Since the 3D feature map of the multiple frames is flattened into a 2D feature map,

these same objects of different scaling are distributed on the same feature map. Therefore, networks need to obtain multi-scale information in the same layer.

3. Feature fusion of the time dimension and the space dimension: Simultaneously performing temporal-spatial feature extraction will result in a heavy computational burden, so most VSR methods first perform temporal feature extraction and then spatial feature extraction. Determining how to effectively integrate the information extracted from time and space has become the main difficulty in the fusion of temporal-spatial features.

The previous VSR research has proposed various methods to overcome these three difficulties. However, none of the other VSR methods use a multi-kernel structure to achieve the diverse receptive field. Most of the VSR methods build deep networks to enlarge the receptive field. The deeper the network, the higher-level feature will be obtained. For super-resolution tasks, high-level information is necessary for the network to identify objects. Nevertheless, low-level information is more critical for the network to capture pixel-wise details. Therefore, our model focuses on obtaining low-level information by deploying parallel channels with multi-kernel structures in the shallow network.

Inspired by the biological vision system, we have designed various multi-convolution kernel structures to tackle the three difficulties. The adaptive multi-kernel structure is used to adjust the receptive field according to the input information. The main benefits of the multi-kernel structure for dealing with the three difficulties are as follows: the adaption to various motions and blurs in motion estimation; the adaption to the scales of each object in multiple frames; the weight deployment of temporal-spatial feature fusion.

For the studies using multi-kernel structure [1, 2, 7] in the other research fields, different convolution kernels are distributed in each layer of the network, and information is fused in each layer and used as the input of the next layer. Unlike these studies, we divide convolution kernels of different sizes into different channels, and different channels exchange information instead of thorough information fusion. The motion patterns of most objects in the same video are similar. We divide the channels of different convolution kernels into main channels and auxiliary channels. The motion of most objects is estimated by the main channel, and the auxiliary channel is only used to expand the receptive field. In addition, we extend the multi-kernel structure to 3D convolution to achieve multiple receptive fields on time-space domain.

In this work, we propose an end-to-end VSR framework using multi-kernel deformable separate 3D convolution. We first describe the framework of our model in Sec. 4.1. We then introduce the four main modules of our model in detail in Sec. 4.2, Sec. 4.3, Sec. 4.4, and Sec. 4.5. Finally, we will show the adjustable multi-kernel structure of our model under different needs in Sec. 4.6.

4.1 Framework of Model

2N+1 consecutive frames are given as the input of our model, and the middle frame is considered as the current frame to be enhanced. The first step of the process is to convert the input frames from RGB to YCbCr color space. Then the three channels of Y, Cb, and Cr are respectively upsampled by bicubic [64] interpolation. To reduce the computational burden, only the Y channel is fed into our model.

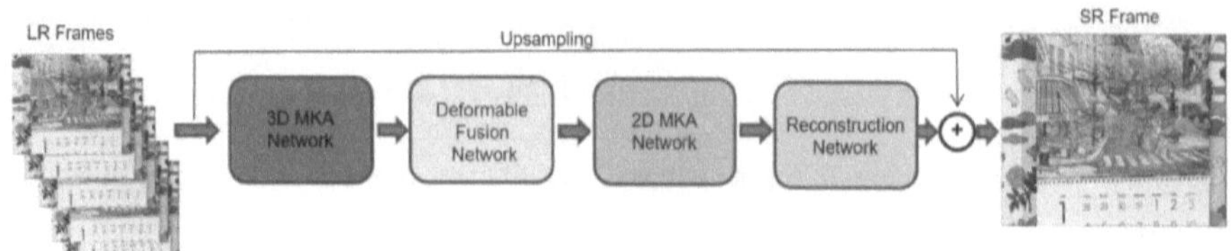

Fig. 4.1 Framework of our proposed VSR method. It consists of four parts: the 3D MKA Network, the MKDC Network, the 2D MKA Network and Reconstruction Network.

Fig. 4.1 illustrates the framework of our proposed method. Our method consists of four trainable networks, including the 3D Multi-Kernel Attention Network (3D MKA Network), the Multi-Kernel Deformable Convolution Network (MKDC Network), the 2D Multi-Kernel Attention Network (2D MKA Network), and the Reconstruction Network.

The purpose of the 3D MKA Network is to extract the temporal-spatial information of input multi-frames and first-stage align the input multiple frames. The 3D convolution can align the different frames according to their spatial information. However, the standard 3D convolution has a weak ability to process the objects in different frames with large deformation and rotation due to its fixed geometric structure. The deformable convolution has better performance in modeling the geometric transformations. We thus design the MKDC Network to further align each pixel of the neighboring frames with the current frame and fuse the feature maps of the multiple input frames. The 3D MKA Network and the MKDC Network jointly complete motion estimation and multi-frame feature fusion operations. Moreover, high-level information is extracted by the 2D MKA Network using deep CNN. Finally, the Reconstruction Network combines the LR feature maps with the interpolated input frame to generate the output HR frame. Multi-kernel methods are deployed in the 3D MKA Network, the MKDC Network, and the 2D MKA Network. And we create the multi-kernel structure according to each module to provide adaptively adjustable receptive fields for the network while avoiding heavy burden.

4.2 3D Multi-Kernel Attention Network

In order to restore more details of the current frame by integrating the information of multiple consecutive frames, 3D multi-kernel convolution method is proposed to adjust the receptive fields in both temporal and spatial domains. However, 3D convolution dramatically increases computational complexity. Tran et al. [39] proposed separate 3D convolution, which can be more easily optimized while achieving similar performance to 3D convolution. Separate 3D convolution is a combination of 1D temporal convolution and 2D spatial convolution. The parameters of the temporal domain and spatial domain can be operated separately. Thus, we employ separate 3D convolution to adjust the receptive fields in the temporal and spatial domains.

Fig. 4.2 demonstrates the architecture of the 3D MKA Network, where CA denotes the channel attention operation, and TA denotes the temporal attention operation. The network in the orange box in Fig. 4.2, which is considered the multi-kernel attention block, repeats five times in the 3D MKA Network. The multi-kernel attention block contains a temporal convolutional layer and a spatial convolutional layer. Each temporal and spatial convolutional layer is composed of different convolutions with kernel sizes of 3 and 5. The channel attention operation aims to fuse the branches of different convolutions in the tem-

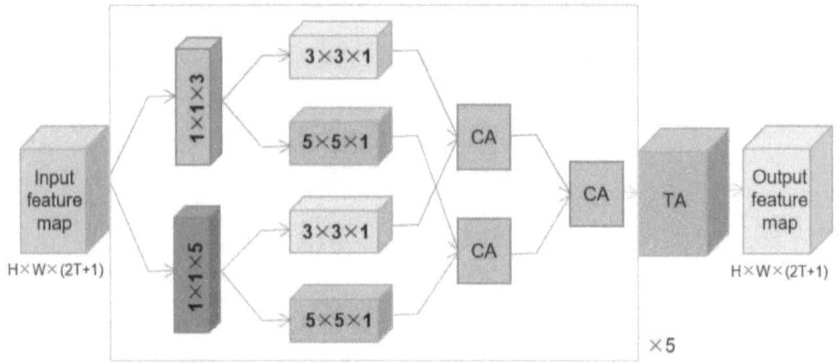

Fig. 4.2 The architecture of the 3D MKA Network.

poral and spatial domain. The multi-kernel attention block can thus be considered as a 3D convolution layer with an adaptive receptive field. The shape of the input feature map is $H \times W \times (2T+1)$. We apply zero-padding to each convolution operation to keep the shape of the output feature map the same as the shape of the input feature map. The output feature map of the multi-kernel attention block is used as the input of the next layer of the multi-kernel attention block. The five repetitions of the multi-kernel attention block can be seen as a five-layer 3D convolution with an adaptive receptive field. Finally, the output feature map of the fifth multi-kernel attention block is used as the input of the temporal attention module.

Let $F_I \in \mathbb{R}^{H \times W \times (2T+1) \times C}$ denote the input feature map of the 3D MKA Network, where W, H, $(2T+1)$, and C represent the width, the height, the number of input frames, and the channel dimension. Let F_{1D}^N denote output feature maps after $1 \times 1 \times N$ convolution operation, then F_{2D}^{MN} denote output feature maps after $M \times M \times 1$ convolution operation.

Let $H_{1D}^N(\cdot)$ and $H_{2D}^M(\cdot)$ denote the convolution operation of kernel size $1 \times 1 \times N$ and $M \times M \times 1$. The separate multi-kernel 3D convolution operation can be described as:

$$F_{1D}^N = H_{1D}^N(F_I), \tag{4.1}$$

$$F_{2D}^{MN} = H_{2D}^M(F_{1D}^N). \tag{4.2}$$

We employ channel attention to assign weights for each channel to fuse the different channels with different receptive fields adaptively. First, we briefly review the channel attention module in SENet [7]. SENet achieves channel attention in two operations: the squeeze operation and the excitation operation. The squeeze operation performs global average pooling for each channel separately. Then, the excitation operation uses two fully connected layers and a gating mechanism with Sigmoid activation to produce channel-wise weights.

We integrate the channel attention to the separate 3D convolution and perform the channel attention fusion on the temporal and spatial convolutions separately. The output feature maps of $1 \times 1 \times N$ convolution and $M \times M \times 1$ convolution are $F_{1D}^N \in \mathbb{R}^{W \times H \times (2T+1) \times C}$ and $F_{2D}^{MN} \in \mathbb{R}^{W \times H \times (2T+1) \times C}$. The weight of temporal channels and spatial channels can be described as:

$$w_t = \frac{1}{2T+1} \sum_{N=3,5} \sum_{i=1}^{2T+1} F_{1D}^N(i), \tag{4.3}$$

$$w_s = \frac{1}{H \times W} \sum_{M=3,5} \sum_{i=1}^{H} \sum_{j=1}^{W} F_{2D}^{MN}(i,j). \tag{4.4}$$

Similar to SEnet, the output of the excitation operation θ is obtained by a gating mechanism:

$$\theta = \delta(W_2 ReLU(W_1 w)), \tag{4.5}$$

where w is the channel's weight obtained by Eqn. (4.3) and Eqn. (4.4). δ is the Sigmoid function. W_1 and W_2 are fully connected operations, the dimension of W_1 is $C \times (\frac{C}{r})$, and the dimension of W_2 is $(\frac{C}{r}) \times C$. The role of reduction ratio r is to limit the complexity, and r is set to 16 by default.

Temporal attention aims to compute the similarity between frames. We apply temporal attention to calculate the similarity between each neighboring frame and the current frame, in order to assign weights to the feature maps of each neighboring frame. The similarity with the feature maps can be expressed by similarity distance D [20], which is expressed as:

$$D = \delta((Conv(F^n)) \odot (Conv(F^c))), \tag{4.6}$$

where F^c is the feature map of the current frame, which is considered as the reference frame, and F^n represents the feature map of the neighboring frame of the current frame. δ is the Sigmoid function. $\odot$ denotes the element-wise multiplication, and $Conv()$ denotes the convolution operation.

Let CA^N and CA^M denote channel attention operation of the time dimension and space dimension. Suppose F_i is the ith channel of the input feature map of the channel attention operation. θ_i denotes one of the components of θ with respect to the ith channel. The channel attention operation $CA(\cdot)$ can be expressed as $CA(F_i) = F_i \cdot \theta_i$. Let F denote the input feature map of the temporal attention operation. The temporal attention operation $TA(\cdot)$ can be expressed as $TA(F) = F \odot D$, where $\odot$ indicates the element-wise multiplication. The output of the multi-kernel attention network F_o can be described as:

$$F_o = \sum_{N=3,5} \sum_{M=3,5} TA(CA^N(CA^M(F_{2D}^{MN}))), \tag{4.7}$$

where $F_o \in \mathbb{R}^{H \times W \times (2T+1) \times C}$ is the output feature map of the 3D MKA Network.

4.3 Feature Fusion with Multi-Kernel Deformable Convolution

Before concatenating the feature maps after 3D convolution, we use deformable convolution to align the feature maps of neighboring frames. Unlike the original deformable convolution, which is utilized to extract features for object detection [5]. We use the deformable convolution to adjust the position of each pixel in the feature map of the neighboring frames by adding an offset Δp_i to each basic convolution. With the receptive field size K, we can have the aligned feature map F_a^n:

$$F_a^n(p_0) = \sum_{i=1}^{K} w(p_i) \cdot F^n(p_0 + p_i + \Delta p_i), \tag{4.8}$$

where p_0 denotes a location on the output feature map F_a^n, p_i enumerates the locations in the sampling grid of basic convolution over the input feature map F^n. $w(p_i)$ denotes the weight for the location p_i.

In the existing SR methods using deformable convolution, two layers of 3×3 convolution are commonly used to obtain the offset vector of each pixel. However, the offset Δp_i is limited by the size of the receptive field. The adaptive receptive field has the potential to generate an offset field that is suitable for various motions. Therefore, we propose a deformable convolution with multiple kernels.

Similarly, we use channel attention to fuse the offset obtained by different convolutional

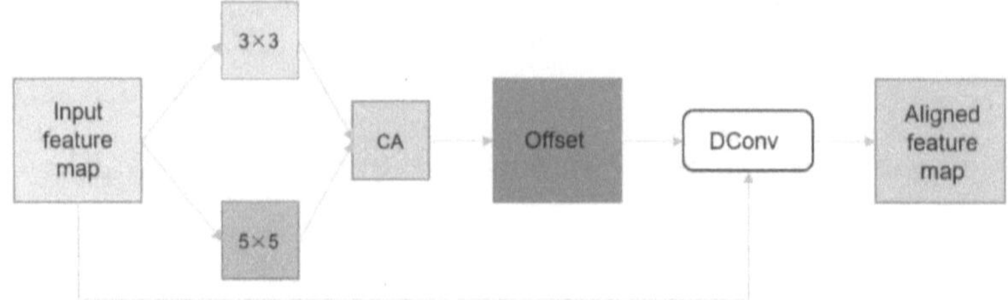

Fig. 4.3 Illustration of multi-kernel deformable convolution.

kernels. Let $F_{offset} = \{\Delta p_i\}$ denote the trainable offset map, and $H^O(\cdot)$ indicate the convolution operation of kernel size O. The convolution kernel used to obtain the offset map is defined with $p_i \in \{(k, l)\}$, where k and l are integers, $k \in [-\frac{O-1}{2}, \frac{O-1}{2}]$ and $l \in [-\frac{O-1}{2}, \frac{O-1}{2}]$. The trainable offset map F_{offset} is obtained by convolution operation $H^O(\cdot)$ with kernel sizes 3 and 5.

$$F_{offset} = CA\left(\sum_{M=3,5} H^O(F^n)\right). \tag{4.9}$$

Fig. 4.3 illustrates the multi-kernel deformable convolution operation. Multi-convolution kernel displacement estimation can be regarded as a sub-channel of the network. The obtained displacement of each pixel is combined by the channel attention module as the offset map.

We concatenate the aligned feature maps along temporal dimension for further 2D convolution. The process of feature fusion with multi-kernel deformable convolution is shown in Fig. 4.4. The output feature map after feature fusion is converted from $F_{o3D} \in \mathbb{R}^{H \times W \times (2T+1) \times C}$ to $F_{o2D} \in \mathbb{R}^{H \times W \times (2T+1)C}$.

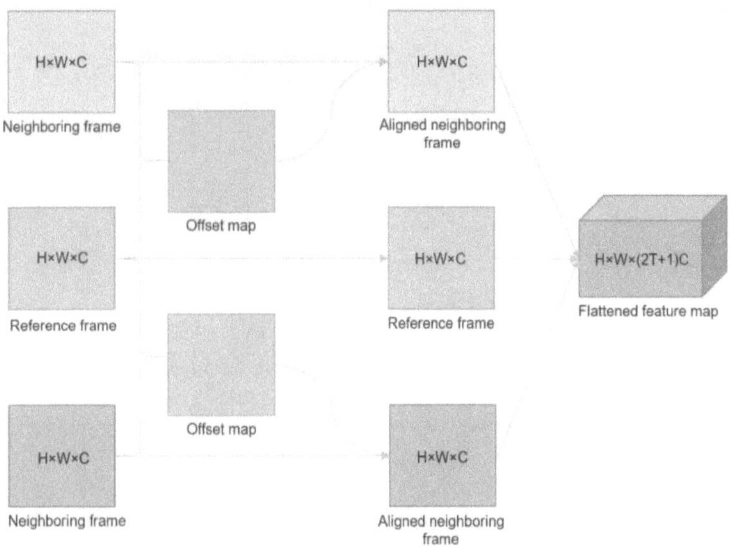

Fig. 4.4 Feature fusion with multi-kernel deformable convolution.

4.4 2D Multi-kernel Attention Network

The goal of the 2D MKA Network is to extract high-level features. As shown in Fig. 4.5, a 40-layer residual structure is applied to the 2D network to obtain high-level features. Large convolution kernels will bring heavy overheads for deep networks. To solve this problem, dilated convolution is used to reduce the amount of computation. Specifically, we use 3×3 convolutions with dilation 2 to approximate the receptive field of a 5×5 convolution kernel, while maintaining the computation the same as 3×3 convolutions. The flattened feature map F^I is divided into two channels corresponding to 3×3 convolution and 5×5 convolution for processing. We regard the channel containing 3×3 convolution as the main channel and the channel with 5×5 convolution as the auxiliary channel.

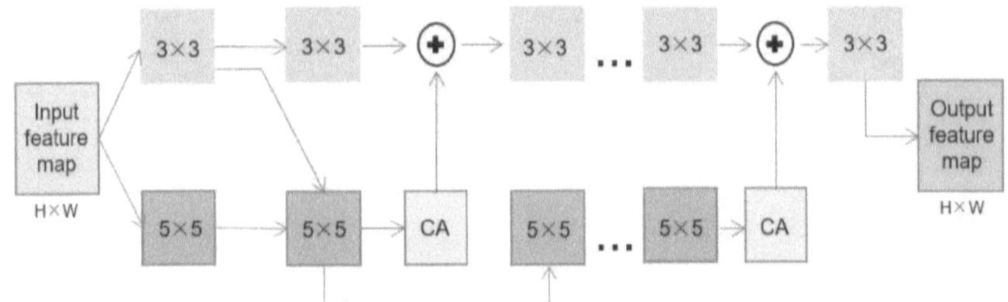

Fig. 4.5 The architecture of the 2D MKA Network.

Let $H^S(\cdot)$ and $H^L(\cdot)$ denote convolution operation with kernel size 3×3 and 5×5. The output feature maps of the first layer are:

$$F_1^S = H^S(F^I), \tag{4.10}$$

$$F_1^L = H^L(F^I), \tag{4.11}$$

where $F^I \in \mathbb{R}^{H \times W \times (2T+1)C}$ is the input feature map of the 2D MKA Network. Then the output feature maps of the second layer are:

$$F_2^L = H^L(F_1^S) + H^L(F_1^L), \tag{4.12}$$

$$F_2^S = H^S(F_1^S) + CA(F_2^L), \tag{4.13}$$

where CA denotes the channel attention module. The output feature map of the channel using 3×3 convolution is considered the final output of the 2D MKA network.

4.5 Reconstruction Network

To restore the low-resolution feature maps to high-resolution frames, we use sub-pixel convolution [10] to upscale the feature maps of multiple frames.

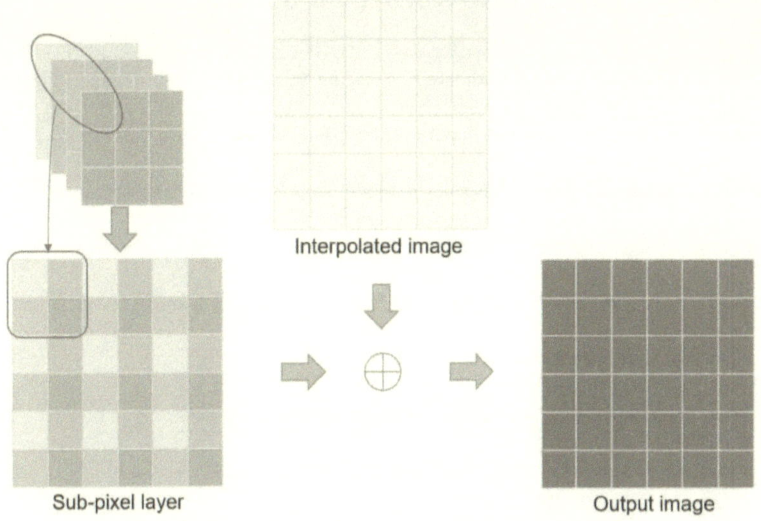

Fig. 4.6 Illustration of of the Reconstruction Network.

The function of sub-pixel convolution is to up-sample feature maps. Unlike other up-sampling methods such as bicubic interpolation and deconvolution, sub-pixel convolution directly obtains up-sampling information from the multiple LR feature maps, avoiding invalid information. To implement the sub-pixel convolution, we use depth-to-space transformation to transfer the feature map's value in the height and width dimensions to the depth dimension. As shown in Fig. 4.6, the pixels of LR feature maps are considered as sub-pixels. These sub-pixels are merged and added to the LR frame after bicubic interpolation to generate the HR frame.

4.6   Adjustable Multi-Kernel Structure

The size of our network can be adjusted according to different requirements. Specifically, our model can be adjusted to the maximum convolution kernel size of 3, 5, and 7. Using a larger maximum kernel size will enable the model to achieve better performance, and it will also require a longer training time. According to different needs, our model can adjust the maximum convolution kernel size in the 3D MKA, the 2D MKA, and the MKDC modules, and the model structure will also be adjusted accordingly. For example, if the input image contains fast object motion and a large amount of blur, more attention should be paid to time information, and a larger 3D MKA module should be used. If the object in the input image moves slowly and contains more details, a larger 2D MKA module can be used to obtain richer spatial information. The larger MKDC module has better performance when the object has apparent rotation and deformation.

The 2D and 3D MKA modules with a maximum convolution kernel size of 3 can be regarded as a standard 2D convolution and standard separate 3D convolution. The 3D convolution module with a maximum convolution kernel of 3 cannot take advantage of the multi-kernel structure, but it can be utilized in fine-tuning.

From Fig. 4.2 to Fig. 4.5, we have illustrated the architecture of the 3D MKA, MKDC, and 2D MKA modules of a maximum kernel size of 5. Next, we introduce the structure of the multi-kernel architectures with a maximum convolution kernel of 7.

With the size of the maximum kernel adjusted, the architecture of the 3D MKA module of maximum kernel size 7 is illustrated in Fig. 4.7. It is worth mentioning that the 3D MKA

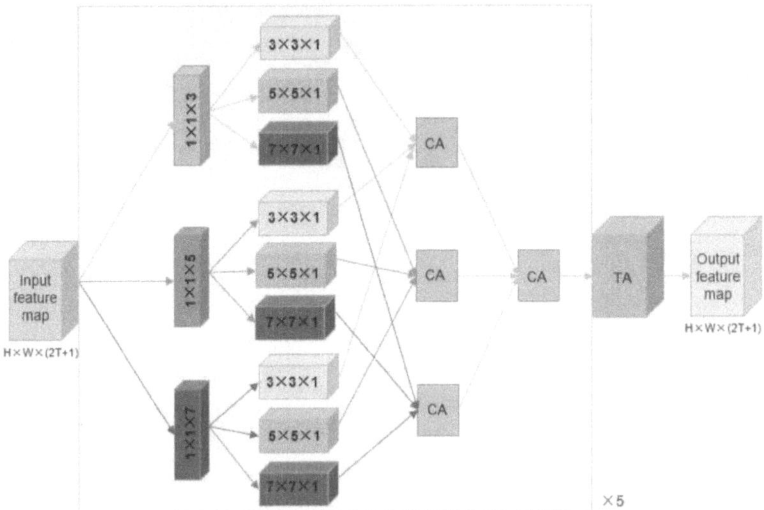

Fig. 4.7 The architecture of the 3D MKA network (maximum kernel size 7).

module with a maximum convolution kernel size of 7 needs to use seven consecutive frames as input. Since we replaced the large convolution kernels with dilated convolutions, using 3D convolutions with the size of 7 will not bring a heavy computational burden.

The architecture of the 2D MKA module of maximum kernel size 7 is illustrated in Fig. 4.8. The channel with the convolution kernel size of 7 is used as the auxiliary channel of the channel with the convolution kernel size of 5. For the 2D MKA module with maximum kernel size 7, let $H^M(\cdot)$ and $H^L(\cdot)$ denote convolution operation with kernel size 5×5 and 7×7. The output feature maps of the first layer will be:

$$F_1^S = H^S(F^I), \tag{4.14}$$

$$F_1^M = H^M(F^I), \tag{4.15}$$

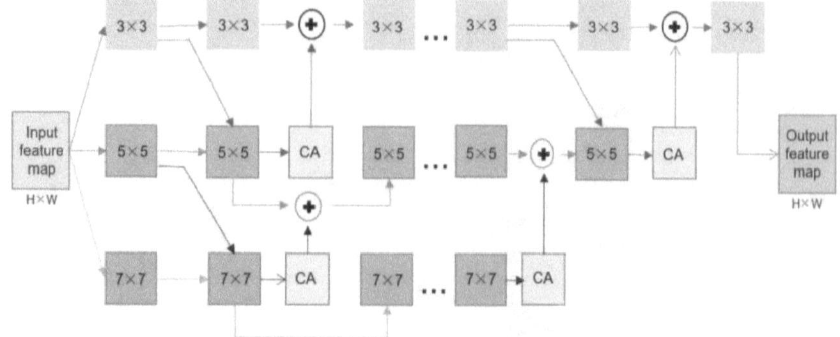

Fig. 4.8 The architecture of the 2D MKA network (maximum kernel size 7).

$$F_1^L = H^L(F^I).$$

(4.16)

And the output feature maps of the second layer will be:

$$F_2^L = H^L(F_1^S) + H^L(F_1^M) + H^L(F_1^L),$$

(4.17)

$$F_2^M = H^M(F_1^M) + CA(F_2^L),$$

(4.18)

$$F_2^S = H^S(F_1^S) + CA(F_2^M).$$

(4.19)

Benefiting from the adjustable multi-convolution kernel structure, our model can be adjusted according to hardware performance and input data.

4.7 Summary

In this chapter, we described the framework of our proposed VSR method. And we presented the significant components in our method, including the 3D MKA Network, the MKDC Network, the 2D MKA Network, and the Reconstruction Network. We described the temporal attention that we employ to assign pixel-level weight to each consecutive frame. We introduced our multi-kernel structure, and we presented the process of integrating the channel attention to the multi-kernel structure to form the multi-kernel attention block. We provided details of applying the multi-kernel attention blocks to separate 3D convolution, deformable convolution, and 2D convolution to form the 3D MKA Network, the MKDC Network, and the 2D MKA Network, respectively. Finally, we presented the adjustable multi-kernel structure to respond to different requirements in VSR tasks.

Chapter 5

Experimental Results

5.1   Environment Setup

The hardware for training and testing our model is a PC equipped with an Intel Core i7-7700K CPU and an NVIDIA Geforce GTX 1080Ti GPU. The PC has 16 GB system memory to store our training and testing dataset and 11 GB GPU-dedicated memory for training our large network.

We use Tensorflow 1.14 [65] as the software platform to implement our method. Tensor-Flow is an open-source software library developed by Google for Deep Neural Networks.

5.1.1   Training Datasets

The correlation between the training set and the test set directly affects the performance of the model. However, there is no standard training dataset for VSR tasks. In our experiment, the mismatch between the training dataset and the test dataset results in a performance degradation of 0.3 dB. The different scales of the dataset also impact the performance of the model. Most VSR methods focus on the up-sampling scales of $\times 2$, $\times 3$, and $\times 4$, among which $\times 4$ is the most challenging up-sampling scale [17, 20, 18], so we set the training set for the up-sampling scale of $\times 4$. We use bicubic downsampling to ensure that the downsampling method is consistent with the test set. Also, because our model uses Bicubic interpolation, the Bicubic downsampling method can reduce the errors caused by sampling methods.

Since our VSR network is larger and deeper than many previous VSR networks, we need a larger dataset to train the network. Therefore, we selected video clips from different datasets to create the large dataset required by our network. In total, 31 of the video clips in our dataset containing 4K, 1080p, and 720p clips were selected from the website of Sajjadi et al. [18], and 20 of them are 1080p video clips selected from `YouTube.com`. The training set contains a total of 74,940 frames. We cropped the image size of each frame into 384×384 to form HR training data. We downsampled these frames by scale ratio 4 to develop training data.

Some VSR methods use a recurrent structure to input data into the network, and other VSR methods generally use three, five, or seven consecutive frames as the input of the model. According to the experimental results in recent VSR works, more consecutive input frames tend to have better test results but also cause a heavier computational burden. Benefiting from the improvement of hardware, VSR studies in recent years have used the largest number of consecutive frames (i.e., seven) as model input [57].

As a summary, Table. 5.1 shows the information of our training dataset.

Table 5.1 Information about the training dataset.

Information	HR resolution	LR resolution	Total frames	Input Size
Training dataset	384×384	96×96	74,940	7

5.1.2 Testing Datasets

Previous VSR papers mostly use the test dataset Vid4 (Videoset4) for comparison, and we also use Vid4 as the main test dataset. VID4 contains four videos corresponding to four different scenes:

1. Calendar contains rich printed text and hand-drawn graphics.

2. The city contains buildings composed of lines with rich details with different scales.

3. Foliage consists of fast-moving vehicles and relatively stationary buildings and trees.

4. Walk contains a large number of human faces and fast-moving pigeons.

Because the VSR models use different methods for data input, the test setup is also different. For example, in the VSR methods used in our comparison, FRVSR [18] uses recurrent frames to read videos. SPMC [17] and VESPCN [16] use three consecutive frames as input; VSRnet [15], TOFlow [66], DUF [19], and EDVR [20] use seven consecutive frames as input. Different numbers of input frames have a non-negligible impact on the test results. We tested three, five, and seven consecutive frames as input. The test results are shown in Sec. 5.5.

5.2　Training Strategy

To generate the input frames, we used bicubic interpolation to increase the image size of LR data to the same size as HR data.

Adam optimizer with $\beta_1 = 0.9$, $\beta_2 = 0.999$ was used for training. Since a large number of residual blocks are used in the network, the initial learning rate was set to 10^{-4}, which is commonly used in residual networks. We used MSE as our loss function. The activation function of the attention module is Sigmoid, and the activation function used for the multi-kernel convolution block and the multi-kernel deformable convolution network is ReLU.

5.3　Evaluation and Loss Function

Peak-signal-to-noise ratio (PSNR) and structural similarity (SSIM) are the two main evaluation metrics of image quality in VSR tasks [67].

PSNR is the most common and widely used objective measurement method for evaluating image quality. It is usually used to measure the image quality after processing to test the performance of the image processing method. The expression of PSNR is defined as:

$$PSNR = 10 Log_{10}\left(\frac{MAX^2}{MSE}\right),\qquad(5.1)$$

where MAX denotes the maximum pixel value in the image. And MSE denotes the mean squared error, which is formulated as:

$$MSE = \frac{1}{MN}\sum_{i=0}^{M-1}\sum_{j=0}^{N-1}[I(i,j) - K(i,j)],\qquad(5.2)$$

where M and N denote the height and width of the image. I and K denote the reference image and the generated image, respectively. The smaller the MSE between the generated image and the original image, the higher the PSNR, which is considered as a higher restoration quality of the image. Therefore, MSE is generally used as the loss function of the SR model.

Although PSNR is the main reference standard for image restoration quality, many experimental results show that the level of PSNR cannot be entirely consistent with the visual quality seen by human eyes. For example, the image restoration quality in contrast and brightness cannot be fully reflected in PSNR. SSIM has been proven to be an image restoration index that is closer to human vision. SSIM mainly compares the contrast, brightness, and structural similarity of the generated image and the original image. SSIM can be expressed as:

$$SSIM(X, Y) = [l(x, y)^\alpha \cdot c(x, y)^\beta \cdot s(x, y)^\gamma], \tag{5.3}$$

where x and y denote the generated image and the reference image. $l(x, y)^\alpha$ represents luminance, $c(x, y)^\beta$ represents contrast, and $s(x, y)^\gamma$ represents structure comparison. α, β, and γ denote three importance factors, and they are set to 1 in our experiment. The comparison measurements can be expressed as:

$$l(x, y) = \frac{2\mu_x\mu_y + c_1}{\mu_x^2 + \mu_y^2 + c_1}, \tag{5.4}$$

$$c(x, y) = \frac{2\sigma_x\sigma_y + c_2}{\sigma_x^2 + \sigma_y^2 + c_2}, \tag{5.5}$$

$$s(x, y) = \frac{2\sigma_{xy} + c_3}{\sigma_x \sigma_y + c_3}, \tag{5.6}$$

where μ_x and μ_y are average values of x and y; $\sigma_x \sigma_y$ is the cross variance of x and y, and σ_x and σ_y are the standard deviations of x and y. By default, $c_3 = 0.5 \cdot c_2$, $c_1 = (k_1 L)^2$, $c_2 = (k_2 L)^2$, $k_1 = 0.01$, and $k_2 = 0.02$, where L is the maximum value of pixels in the image, the value of L is usually 255 in gray and color images.

Both PSNR and SSIM are calculated only in the luminance channel, and the higher the PSNR and SSIM values, the better the image restoration quality.

5.4 Color Space

As mentioned in Sec. 4.1, we first convert the image from RGB color space to YCbCr color space before inputting the LR image to the network, and we only utilize the Y (luminance) channel for training and testing.

RGB and YCbCr are two common color spaces for color representation. RGB is a human perception-based model to reproduce various colors by adding the red, green, and blue channels together in various ways. However, the three channels of the RGB color space contain hue, brightness, and saturation at the same time. So the model using the RGB color space must be trained separately for the three channels, which causes a heavy computational burden. Some SISR methods with a low computational complexity directly use the RGB color space for training and testing. YCbCr color space is defined by coordinate transformation through associated RGB primaries and white points. In YCbCr, Y refers to the luma component, Cb refers to the blue-difference chroma component, and Cr refers to the red-difference chroma component. Human vision is sensitive to changes in brightness but not to changes in chrominance components. Therefore, most VSR methods only use the Y channel in the YCbCr color space for training and testing.

ITU-R BT.601 is used as the conversion standard for RGB and YCbCr color space in MATLAB. The YCbCr information is transformed from the corresponding RGB information by the following equation:

$$Y = 16 + \frac{65.738 \cdot R}{255} + \frac{129.057 \cdot G}{255} + \frac{25.064 \cdot B}{255} \tag{5.7}$$

$$Cb = 128 - \frac{37.945 \cdot R}{255} - \frac{74.494 \cdot G}{255} + \frac{122.439 \cdot B}{255} \tag{5.8}$$

$$Cr = 128 + \frac{112.439 \cdot R}{255} - \frac{94.154 \cdot G}{255} - \frac{18.285 \cdot B}{255} \tag{5.9}$$

JEPG is another standard transformation for RGB and YCbCr color space, which is implemented in OpenCV (the default image-reading color space in OpenCV is GBR) and other Python libraries, such as Spicy and PIL (Python Imaging Library). The transformation equation of JEPG transformation from RGB to YCbCr is demonstrated by the following equation:

$$Y = 0 + 0.299 \cdot R + 0.587 \cdot G + 0.114 \cdot B \tag{5.10}$$

$$Cb = 128 - 0.16876 \cdot R - 0.331264 \cdot G + 0.5 \cdot B \tag{5.11}$$

$$Cr = 128 + 0.5 \cdot R - 0.418658 \cdot G - 0.081312 \cdot B \qquad (5.12)$$

In our experiment, LR images were read by Python-Opencv and converted to YCbCr color space because the model is trained in Python environment. We first use bicubic interpolation to upsample the three input channels, Y, Cb, and Cr, respectively. We then separately train the Y channel and combine the trained Y channel with the up-sampled Cb and Cr channels to form an output HR image. Finally, we use MATLAB to perform PSNR testing on the Y channel of the output HR images.

5.5   Experimental Results

We first tested the performance of our method on the Vid4 test set. Since our model adjusts the structure according to the numbers of input frames, the difference in the number of input frames has a great influence on the results of our model. Table 5.2 shows the test result of our method using different numbers of input frames on Vid4. In order to achieve the same comparison conditions as the majority of recent VSR methods, which use seven frames as input, we also used seven frames as the input of the model in the subsequent experimental comparison.

In Sec. 5.5.1, we compare the test result of our model on Vid4 with the previous state-of-the-art methods. Moreover, in Sec. 5.5.2, we build a comparison model to undertake the ablation studies to prove the effectiveness of each part of our model.

Table 5.2 Quantitative result of our method using different numbers of input frames
on Vid4 for ×4 video SR.

Input Frames		Bicubic	3 Frames	5 Frames	7 Frames
Calendar	PSNR	20.39	23.70	24.18	24.32
	SSIM	0.572	0.786	0.812	0.822
City	PSNR	25.16	28.15	28.68	28.78
	SSIM	0.603	0.797	0.827	0.842
Foliage	PSNR	23.47	26.01	26.26	26.40
	SSIM	0.567	0.731	0.767	0.781
Walk	PSNR	26.10	30.26	30.63	30.68
	SSIM	0.797	0.891	0.910	0.916
Average	PSNR	23.78	27.03	27.43	27.55
	SSIM	0.635	0.801	0.829	0.840

5.5.1 Comparison with Other VSR Methods

Using different training sets has a great impact on the test results. In our experiment,
the mismatch between the training dataset and the test dataset resulted in a performance
degradation of 0.3 dB. However, most of the existing VSR methods use various other training datasets, which makes it difficult to achieve a fair comparison between different VSR
methods.

We compare our proposed method with four other recent VSR methods on the Vid4 test
dataset. TOFlow [66] proposes a trainable motion estimation component and takes seven

consecutive frames as the input of the model. FRVSR [18] proposes a frame-recurrent method for the input of the model. Due to its recurrent nature, FRVSR has the ability to assimilate previous frames without increasing computational parameters. DUF [19] generates dynamic upsampling filters and residual images to avoid explicit motion compensation and uses seven frames as model input. EDVR [20] designed a pyramidal learning structure, using deformable convolution and attention mechanism for motion compensation and using seven frames as the input of the model. EDVR is considered to be the latest best-performing VSR method.

Table 5.3 Quantitative comparison on Vid4 for $\times 4$ video SR.

Method		*TOFlow*	*FRVSR*	*DUF-52L*	*EDVR*	*Our Result*
Calendar	PSNR	22.47	-	24.04	24.25	24.32
	SSIM	0.732	-	0.811	0.815	0.822
City	PSNR	26.78	-	28.27	28.00	28.78
	SSIM	0.740	-	0.831	0.812	0.842
Foliage	PSNR	25.27	-	26.22	26.34	26.40
	SSIM	0.709	-	0.757	0.764	0.781
Walk	PSNR	29.05	-	30.70	31.02	30.68
	SSIM	0.879	-	0.909	0.915	0.916
Average	PSNR	25.89	26.69	27.33	27.35	27.55
	SSIM	0.765	0.822	0.831	0.826	0.840

Table 5.3 illustrates the quantitative comparison results. On Vid4, our method achieved the highest PSNR and SSIM compared to previous VSR methods. Specifically, our proposed

method outperformed the previous state-of-the-art methods with an average PSNR of 0.2 dB and an average SSIM of 0.014.

Fig. 5.1 demonstrates the qualitative comparison of our proposed method with the other four VSR methods on Vid4-Calendar. It can be observed that our proposed model is able to restore fuzzy textures. Fig. 5.2 shows the visual comparison on Vid4-City. It can be observed that our method can restore the outline of the windows more clearly than the other methods. Fig. 5.3 illustrates that our method performs better in the test to restore high-speed vehicles. For example, it can restore the rims, auxiliary mirrors, and shadows of the vehicle more visually realistically than other methods. In Fig. 5.4, it can be observed that the human's faces are deformed and blurred in the test results of other methods and that our method avoids this deformation and blur.

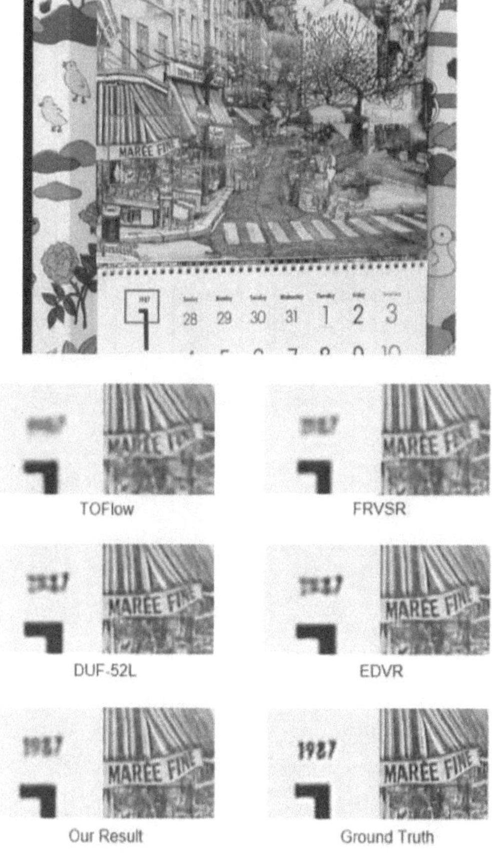

Fig. 5.1 Visual comparison on Vid4-Calendar.

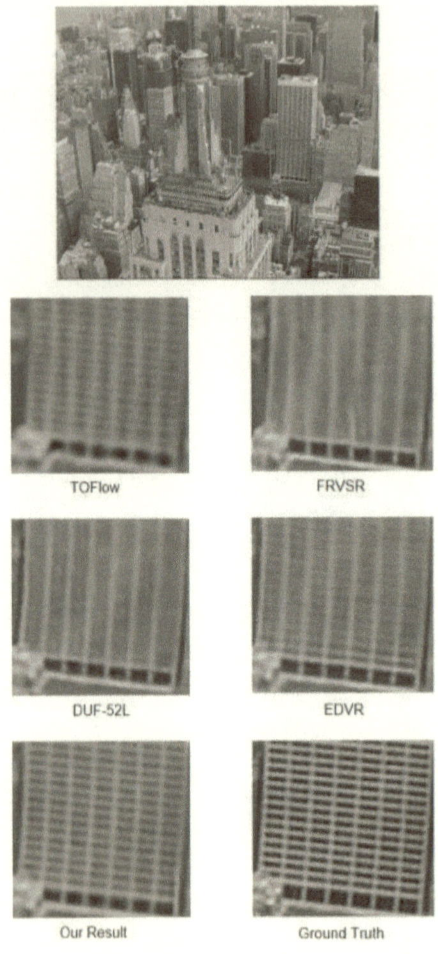

Fig. 5.2 Visual comparison on Vid4-City.

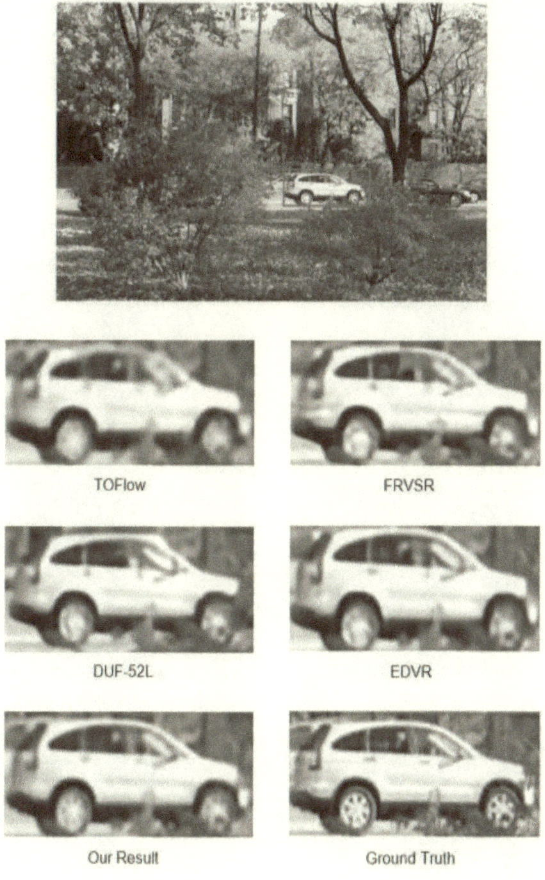

Fig. 5.3 Visual comparison on Vid4-Foliage.

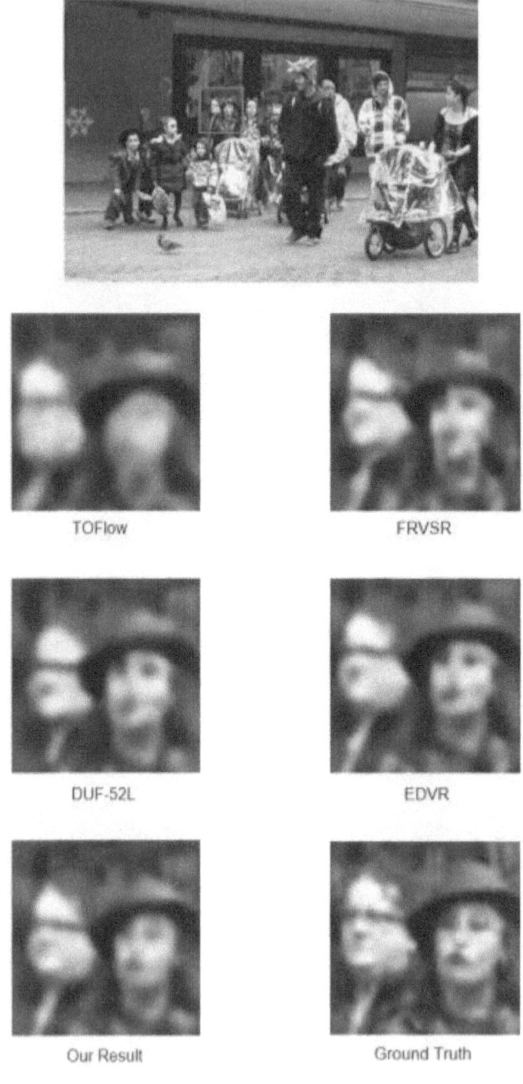

Fig. 5.4 Visual comparison on Vid4-Walk.

5.5.2 Ablation Studies

In order to demonstrate the effectiveness of each module in our work, we created three ablation models. Model A only employs convolution kernels of size 3 as the baseline model. Model B adopts 3D multi-kernel blocks and standard 2D convolution layers. Model C adopts standard 3D convolution layers and 2D multi-kernel blocks. These comparison models have similar computation parameters to our proposed model. We also demonstrate the effectiveness of multi-kernel offset acquisition for deformable convolution by creating Model D. Model D is equipped with the 3D MKA module and 2D MKA module but uses regular deformable convolution for alignment.

Table 5.4 Ablation studies result on Vid4 for $\times 4$ video SR.

Method Name	Model A	Model B	Model C	Model D	Our Model
PSNR	26.63	27.09	27.33	27.43	27.54
SSIM	0.809	0.827	0.831	0.842	0.839

The quantitative results are shown in Table 5.4. Our model achieves better PSNR test results compared to the comparison models. It is worth noting that the output of Model D has a lower PSNR than the output of our model, but has a higher SSIM than the output of our model. This situation also appeared in previous VSR studies. For example, the PSNR of EDVR is higher than that of DUF-52L, but SSIM is lower than DUF-52L. Since most VSR studies use MSE as the loss function, MSE is directly related to PSNR but not directly related to SSIM. The model with higher PSNR output results is better optimized and has a more vital ability to process unknown samples.

The visual comparisons of the comparison models and our proposed model are shown from Fig. 5.5 to Fig. 5.8. We can see from Fig. 5.5 and Fig. 5.6 that our model can restore printed text and texture details of the building more clearly than the comparison models. For Fig. 5.7, it can be observed in the areas marked by red boxes that our result is more visually close to the ground truth than the ablation models. Furthermore, the test result for human faces is illustrated in Fig. 5.8. It is evident that Model B, which uses the 3D MKA module, is able to restore more visually real faces, while Model C, which adopts the 2D MKA module, performs better at restoring edges and textures than the baseline model. Our research thus demonstrates that 2D and 3D MKA modules and MKDC modules result in clearer and more realistic outputs than the comparison models.

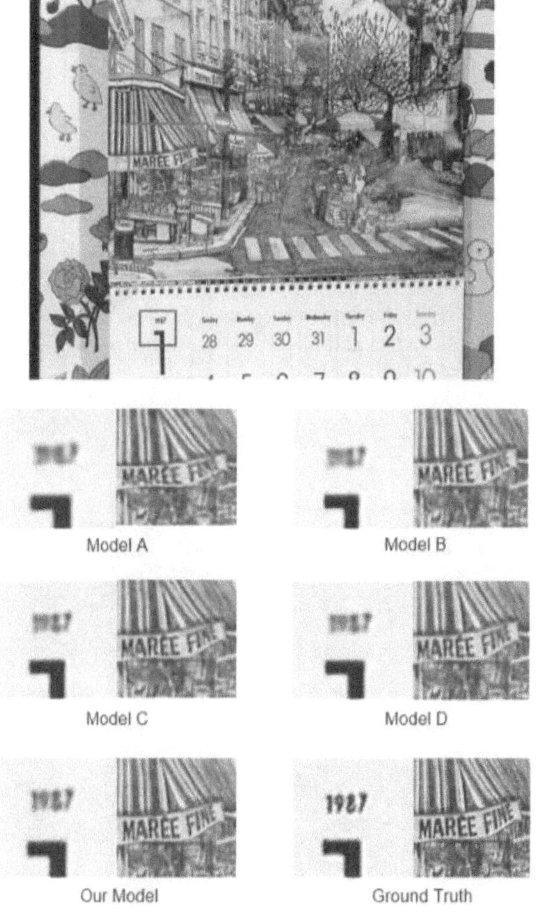

Fig. 5.5 Visual comparison of ablation studies (Calendar).

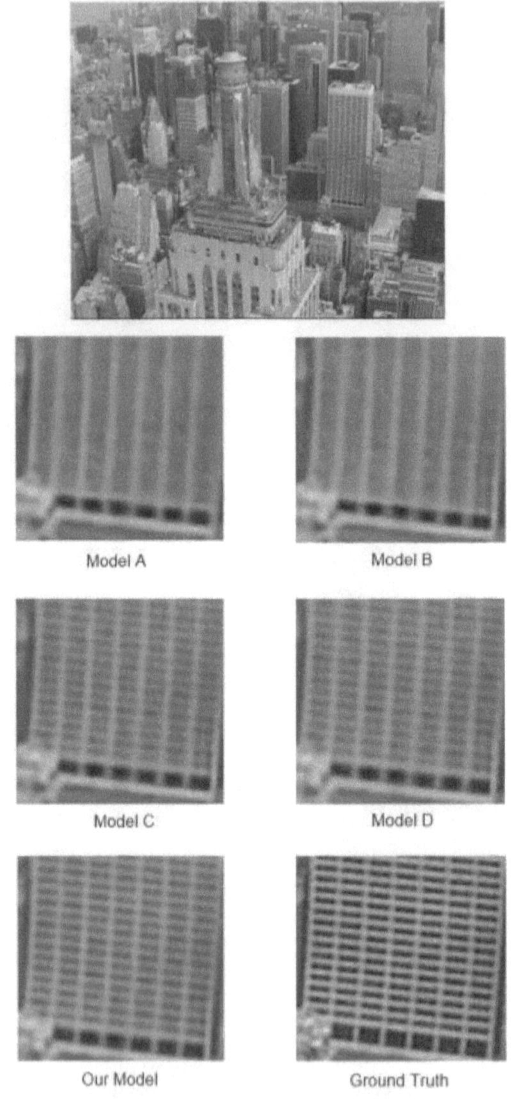

Fig. 5.6 Visual comparison of ablation studies (City).

Fig. 5.7 Visual comparison of ablation studies (Foliage).

Fig. **5.8** Visual comparison of ablation studies (Walk).

5.5.3   Evaluation on Additional Examples

To verify the effectiveness of our method on additional examples, we use the additional
dataset provided by Tao et al. [17] for testing. Their test dataset is published on their
website. The test data set contains 30 videos, and each video contains 31 frames. They
proposed these videos to provide more comprehensive performance tests of their method
SPMC.

The visual results of additional examples are shown in Fig. 5.9 to Fig. 5.12. For
comparison, these images contain the test results of our method and the test results of
SPMC. The PSNR and SSIM of each test sequence are also shown in these figures. Our
method achieves higher PSNR and SSIM than SPMC for each test sequence. From Fig. 5.5
and Fig. 5.6, we can see that our method can restore richer details of the statues. For Fig.
5.11, our method more realistically restores the shadows and textures of the stairs. From
Fig. 5.12, it can be observed that our method restores the human nose and mouth, which
are not clearly restored in the SPMC's results.

Fig. 5.9 Visual comparison on RMVTG-011.

Fig. 5.10 Visual comparison on HKVGT-004.

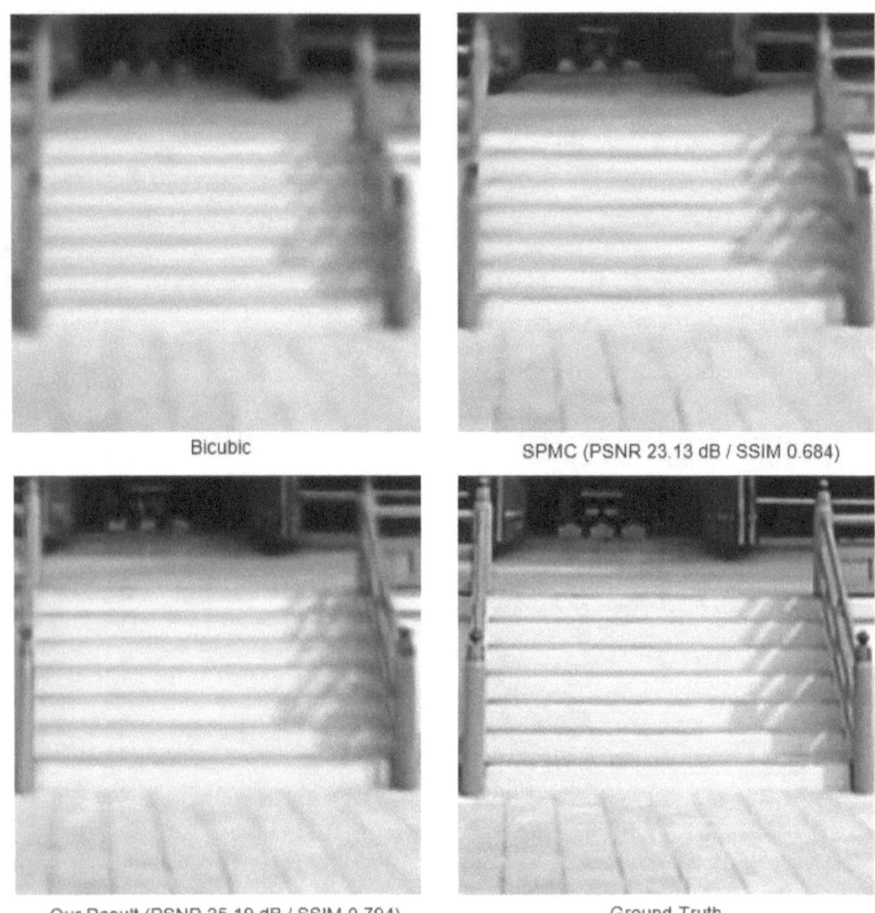

Fig. 5.11 Visual comparison on JVC-009.

91

Fig. 5.12 Visual comparison on VENI3-011.

5.6 Summary

In this chapter, we have described the details of the setup of our experiment environment. Moreover, we have included our training strategy, including evaluation, loss function, and color space. We have presented our experimental results, and we compare our results with the results of other VSR methods. The comparison shows that our proposed method outperforms the other existing VSR methods. We have also demonstrated the effectiveness of each module in our model by ablation studies. Furthermore, we have evaluated our model on additional examples to enrich our experiment.

Chapter 6

Conclusions

In this book, we first introduced the related concepts of CNN and reviewed some SISR and VSR methods. Next, we pointed out the limitations of the CNN with the fixed-size convolution kernel in the VSR task and proposed multi-kernel convolution blocks. Then, we have presented an end-to-end video super-resolution framework using multi-kernel deformable separate 3D convolution. We integrated channel attention to separate 3D convolution to design a 3D multi-kernel attention block. We also proposed a multi-kernel deformable convolution module for multi-frame alignment. We further designed a multi-kernel 2D attention structure to extract high-level features. The ablation experiment validates the effectiveness of each module. The results of the comparison experiment demonstrate that our proposed method outperforms the existing methods due to the 3D MKA module, the MKDC alignment module, and the 2D MKA module.

In the future, we will try to use the frame-recurrent method for data input. We utilized multiple consecutive frames as input. This input method has two main weaknesses. First, the same frame will be processed multiple times, which increases the computational burden. Second, every output frame is generated independently, which limits the temporally consistent. The frame-recurrent method could overcome these two weaknesses by using the previous output to estimate the subsequent frames. However, combining the frame-recurrent method and deformable convolution will cause the model to be difficult to converge, and artifacts may appear in the output. We will try to optimize the fusion method of frame-recurrent method and deformable convolution to enhance our multi-kernel deformable 3D convolution.

To better imitate biological vision and improve the VSR model's performance further, we will try to incorporate 3D image super-resolution technology into our model in the future. For biological visual channels, the receptive fields of some visual cells are distributed on one retina, and the receptive fields of the other visual cells are distributed on two retinas. The visual cells with receptive field distributing on two retains are considered binocular visual cells, which are rarely considered by super-resolution research. To imitate the binocular visual cells, we will use the videos captured by dual cameras as the input. According to different inputs, the receptive field of the binocular visual cells will be adjusted between the information obtained by the right eye and left eye. To simulate this phenomenon, we will use channel attention to combine the video captured by the left camera and the right camera to form a trainable 3D image super-resolution network. Finally, we will merge this network into our proposed model. With the information provided by dual cameras, we aim to enable our model to process 4D information in temporal, spatial, and depth domains.

References

[1] Christian Szegedy, Wei Liu, Yangqing Jia, Pierre Sermanet, Scott Reed, Dragomir Anguelov, Dumitru Erhan, Vincent Vanhoucke, and Andrew Rabinovich. Going deeper with convolutions. In *Proceedings of the IEEE Conference on Computer Vision and Pattern Recognition*, pages 1–9, 2015.

[2] Xiang Li, Wenhai Wang, Xiaolin Hu, and Jian Yang. Selective kernel networks. In *Proceedings of the IEEE/CVF Conference on Computer Vision and Pattern Recognition*, pages 510–519, 2019.

[3] Kaiming He, Xiangyu Zhang, Shaoqing Ren, and Jian Sun. Deep residual learning for image recognition. In *Proceedings of the IEEE Conference on Computer Vision and Pattern Recognition*, pages 770–778, 2016.

[4] Fisher Yu and Vladlen Koltun. Multi-scale context aggregation by dilated convolutions. *arXiv preprint arXiv:1511.07122*, 2015. (Last accessed: 14.02.2021).

[5] Jifeng Dai, Haozhi Qi, Yuwen Xiong, Yi Li, Guodong Zhang, Han Hu, and Yichen Wei. Deformable convolutional networks. In *Proceedings of the IEEE International Conference on Computer Vision*, pages 764–773, 2017.

[6] Max Jaderberg, Karen Simonyan, Andrew Zisserman, and Koray Kavukcuoglu. Spatial transformer networks. *arXiv preprint arXiv:1506.02025*, 2015. (Last accessed: 06.03.2021).

[7] Jie Hu, Li Shen, and Gang Sun. Squeeze-and-excitation networks. In *Proceedings of the IEEE Conference on Computer Vision and Pattern Recognition*, pages 7132–7141, 2018.

[8] Chao Dong, Chen Change Loy, Kaiming He, and Xiaoou Tang. Image super-resolution using deep convolutional networks. *IEEE Transactions on Pattern Analysis and Machine Intelligence*, 38(2):295–307, 2015.

[9] Chao Dong, Chen Change Loy, and Xiaoou Tang. Accelerating the super-resolution convolutional neural network. In *European Conference on Computer Vision*, pages 391–407. Springer, 2016.

[10] Wenzhe Shi, Jose Caballero, Ferenc Huszár, Johannes Totz, Andrew P Aitken, Rob Bishop, Daniel Rueckert, and Zehan Wang. Real-time single image and video super-resolution using an efficient sub-pixel convolutional neural network. In *Proceedings of the IEEE Conference on Computer Vision and Pattern Recognition*, pages 1874–1883, 2016.

[11] Jiwon Kim, Jung Kwon Lee, and Kyoung Mu Lee. Accurate image super-resolution using very deep convolutional networks. In *Proceedings of the IEEE Conference on Computer Vision and Pattern Recognition*, pages 1646–1654, 2016.

[12] Xiao-Jiao Mao, Chunhua Shen, and Yu-Bin Yang. Image restoration using convolutional autoencoders with symmetric skip connections. *arXiv preprint arXiv:1606.08921*, 2016. (Last accessed: 20.03.2021).

[13] Yulun Zhang, Kunpeng Li, Kai Li, Lichen Wang, Bineng Zhong, and Yun Fu. Image super-resolution using very deep residual channel attention networks. In *Proceedings of the European Conference on Computer Vision (ECCV)*, pages 286–301, 2018.

[14] Christian Ledig, Lucas Theis, Ferenc Huszár, Jose Caballero, Andrew Cunningham, Alejandro Acosta, Andrew Aitken, Alykhan Tejani, Johannes Totz, Zehan Wang, et al. Photo-realistic single image super-resolution using a generative adversarial network. In *Proceedings of the IEEE Conference on Computer Vision and Pattern Recognition*, pages 4681–4690, 2017.

[15] Armin Kappeler, Seunghwan Yoo, Qiqin Dai, and Aggelos K Katsaggelos. Video super-resolution with convolutional neural networks. *IEEE Transactions on Computational Imaging*, 2(2):109–122, 2016.

[16] Jose Caballero, Christian Ledig, Andrew Aitken, Alejandro Acosta, Johannes Totz, Zehan Wang, and Wenzhe Shi. Real-time video super-resolution with spatio-temporal networks and motion compensation. In *Proceedings of the IEEE Conference on Computer Vision and Pattern Recognition*, pages 4778–4787, 2017.

[17] Xin Tao, Hongyun Gao, Renjie Liao, Jue Wang, and Jiaya Jia. Detail-revealing deep video super-resolution. In *Proceedings of the IEEE International Conference on Computer Vision*, pages 4472–4480, 2017.

[18] Mehdi SM Sajjadi, Raviteja Vemulapalli, and Matthew Brown. Frame-recurrent video super-resolution. In *Proceedings of the IEEE Conference on Computer Vision and Pattern Recognition*, pages 6626–6634, 2018.

[19] Younghyun Jo, Seoung Wug Oh, Jaeyeon Kang, and Seon Joo Kim. Deep video super-resolution network using dynamic upsampling filters without explicit motion compensation. In *Proceedings of the IEEE Conference on Computer Vision and Pattern Recognition*, pages 3224–3232, 2018.

[20] Xintao Wang, Kelvin CK Chan, Ke Yu, Chao Dong, and Chen Change Loy. Edvr: Video restoration with enhanced deformable convolutional networks. In *Proceedings of the IEEE/CVF Conference on Computer Vision and Pattern Recognition Workshops*, pages 0–0, 2019.

[21] John A Kennedy, Ora Israel, Alex Frenkel, Rachel Bar-Shalom, and Haim Azhari. Super-resolution in pet imaging. *IEEE Transactions on Medical Imaging*, 25(2):137–147, 2006.

[22] Yun Zhang. Problems in the fusion of commercial high-resolution satelitte as well as landsat 7 images and initial solutions. *International Archives of Photogrammetry Remote Sensing and Spatial Information Sciences*, 34(4):587–592, 2002.

[23] Philip J Davis. *Interpolation and approximation*. Courier Corporation, 1975.

[24] Jose Caballero, Christian Ledig, Andrew Aitken, Alejandro Acosta, Johannes Totz, Zehan Wang, and Wenzhe Shi. Real-time video super-resolution with spatio-temporal networks and motion compensation. In *Proceedings of the IEEE Conference on Computer Vision and Pattern Recognition*, pages 4778–4787, 2017.

[25] Kamal Nasrollahi and Thomas B Moeslund. Super-resolution: a comprehensive survey. *Machine Vision and Applications*, 25(6):1423–1468, 2014.

[26] Younghyun Jo, Seoung Wug Oh, Jaeyeon Kang, and Seon Joo Kim. Deep video super-resolution network using dynamic upsampling filters without explicit motion compensation. In *Proceedings of the IEEE Conference on Computer Vision and Pattern Recognition*, pages 3224–3232, 2018.

[27] Ding Liu, Zhaowen Wang, Yuchen Fan, Xianming Liu, Zhangyang Wang, Shiyu Chang, and Thomas Huang. Robust video super-resolution with learned temporal dynamics. In *Proceedings of the IEEE International Conference on Computer Vision*, pages 2507–2515, 2017.

[28] David H Hubel and Torsten N Wiesel. Receptive fields and functional architecture of monkey striate cortex. *The Journal of Physiology*, 195(1):215–243, 1968.

[29] Stephen W Kuffler. Discharge patterns and functional organization of mammalian retina. *Journal of neurophysiology*, 16(1):37–68, 1953.

[30] Robert W Rodieck and Jonathan Stone. Analysis of receptive fields of cat retinal ganglion cells. *Journal of neurophysiology*, 28(5):833–849, 1965.

[31] Jiuxiang Gu, Zhenhua Wang, Jason Kuen, Lianyang Ma, Amir Shahroudy, Bing Shuai, Ting Liu, Xingxing Wang, Gang Wang, Jianfei Cai, et al. Recent advances in convolutional neural networks. *Pattern Recognition*, 77:354–377, 2018.

[32] Jacob Devlin, Ming-Wei Chang, Kenton Lee, and Kristina Toutanova. Bert: Pre-training of deep bidirectional transformers for language understanding. *arXiv preprint arXiv:1810.04805*, 2018. (Last accessed: 04.02.2021).

[33] Sebastian Thrun, Lawrence K Saul, and Bernhard Schölkopf. *Advances in Neural Information Processing Systems 16: Proceedings of the 2003 Conference*, volume 16. MIT press, 2004.

[34] Jiuxiang Gu, Zhenhua Wang, Jason Kuen, Lianyang Ma, Amir Shahroudy, Bing Shuai, Ting Liu, Xingxing Wang, Gang Wang, Jianfei Cai, et al. Recent advances in convolutional neural networks. *Pattern Recognition*, 77:354–377, 2018.

[35] David H Hubel and Torsten N Wiesel. Receptive fields, binocular interaction and functional architecture in the cat's visual cortex. *The Journal of Physiology*, 160(1):106–154, 1962.

[36] Matthew D Zeiler and Rob Fergus. Visualizing and understanding convolutional networks. In *European Conference on Computer Vision*, pages 818–833. Springer, 2014.

[37] Jost Tobias Springenberg, Alexey Dosovitskiy, Thomas Brox, and Martin Riedmiller. Striving for simplicity: The all convolutional net. *arXiv preprint arXiv:1412.6806*, 2014. (Last accessed: 25.04.2021).

[38] Shuiwang Ji, Wei Xu, Ming Yang, and Kai Yu. 3d convolutional neural networks for human action recognition. *IEEE Transactions on Pattern Analysis and Machine Intelligence*, 35(1):221–231, 2012.

[39] Du Tran, Lubomir Bourdev, Rob Fergus, Lorenzo Torresani, and Manohar Paluri. Learning spatiotemporal features with 3d convolutional networks. In *Proceedings of the IEEE International Conference on Computer Vision*, pages 4489–4497, 2015.

[40] Jun Han and Claudio Moraga. The influence of the sigmoid function parameters on the speed of backpropagation learning. In *International Workshop on Artificial Neural Networks*, pages 195–201. Springer, 1995.

[41] Peter Dayan and Laurence F Abbott. *Theoretical neuroscience: computational and mathematical modeling of neural systems*. Computational Neuroscience Series, 2001.

[42] Vinod Nair and Geoffrey E Hinton. Rectified linear units improve restricted boltzmann machines. In *Icml*, 2010.

[43] Djork-Arné Clevert, Thomas Unterthiner, and Sepp Hochreiter. Fast and accurate deep network learning by exponential linear units (elus). *arXiv preprint arXiv:1511.07289*, 2015. (Last accessed: 25.04.2021).

[44] Xavier Glorot and Yoshua Bengio. Understanding the difficulty of training deep feedforward neural networks. In *Proceedings of the thirteenth international conference on artificial intelligence and statistics*, pages 249–256. JMLR Workshop and Conference Proceedings, 2010.

[45] Kaiming He, Xiangyu Zhang, Shaoqing Ren, and Jian Sun. Delving deep into rectifiers: Surpassing human-level performance on imagenet classification. In *Proceedings of the IEEE International Conference on Computer Vision*, pages 1026–1034, 2015.

[46] Sergey Ioffe and Christian Szegedy. Batch normalization: Accelerating deep network training by reducing internal covariate shift. In *International Conference on Machine Learning*, pages 448–456. PMLR, 2015.

[47] Johan Bjorck, Carla Gomes, Bart Selman, and Kilian Q Weinberger. Understanding batch normalization. *arXiv preprint arXiv:1806.02375*, 2018. (Last accessed: 20.12.2020).

[48] Shibani Santurkar, Dimitris Tsipras, Andrew Ilyas, and Aleksander Madry. How does batch normalization help optimization? *arXiv preprint arXiv:1805.11604 (Last accessed: 02.05.2021)*, 2018.

[49] Robert Hecht-Nielsen. Theory of the backpropagation neural network. In *Neural networks for perception*, pages 65–93. Elsevier, 1992.

[50] David E Rumelhart, Geoffrey E Hinton, and Ronald J Williams. Learning representations by back-propagating errors. *Nature*, 323(6088):533–536, 1986.

[51] Léon Bottou. Large-scale machine learning with stochastic gradient descent. In *Proceedings of COMPSTAT'2010*, pages 177–186. Springer, 2010.

[52] Ilya Sutskever, James Martens, George Dahl, and Geoffrey Hinton. On the importance of initialization and momentum in deep learning. In *International Conference on Machine Learning*, pages 1139–1147. PMLR, 2013.

[53] John Duchi, Elad Hazan, and Yoram Singer. Adaptive subgradient methods for online learning and stochastic optimization. *Journal of Machine Learning Research*, 12(7), 2011.

[54] Diederik P Kingma and Jimmy Ba. Adam: A method for stochastic optimization. *arXiv preprint arXiv:1412.6980*, 2014. (Last accessed: 03.02.2021).

[55] Bo Wang, Yang Lei, Sibo Tian, Tonghe Wang, Yingzi Liu, Pretesh Patel, Ashesh B Jani, Hui Mao, Walter J Curran, Tian Liu, et al. Deeply supervised 3d fully convolutional networks with group dilated convolution for automatic mri prostate segmentation. *Medical physics*, 46(4):1707–1718, 2019.

[56] Ryuhei Hamaguchi, Aito Fujita, Keisuke Nemoto, Tomoyuki Imaizumi, and Shuhei Hikosaka. Effective use of dilated convolutions for segmenting small object instances in remote sensing imagery. In *2018 IEEE Winter Conference on Applications of Computer Vision (WACV)*, pages 1442–1450. IEEE, 2018.

[57] Xinyi Ying, Longguang Wang, Yingqian Wang, Weidong Sheng, Wei An, and Yulan Guo. Deformable 3d convolution for video super-resolution. *IEEE Signal Processing Letters*, 27:1500–1504, 2020.

[58] Jun Li, Xianglong Liu, Mingyuan Zhang, and Deqing Wang. Spatio-temporal deformable 3d convnets with attention for action recognition. *Pattern Recognition*, 98:107037, 2020.

[59] Sanghyun Woo, Jongchan Park, Joon-Young Lee, and In So Kweon. Cbam: Convolutional block attention module. In *Proceedings of the European Conference on Computer Vision (ECCV)*, pages 3–19, 2018.

[60] Karen Simonyan and Andrew Zisserman. Very deep convolutional networks for large-scale image recognition. *arXiv preprint arXiv:1409.1556 (Last accessed: 12.19.2020)*, 2014.

[61] Matthew D Zeiler, Graham W Taylor, and Rob Fergus. Adaptive deconvolutional networks for mid and high level feature learning. In *2011 International Conference on Computer Vision*, pages 2018–2025. IEEE, 2011.

[62] Justin Johnson, Alexandre Alahi, and Li Fei-Fei. Perceptual losses for real-time style transfer and super-resolution. In *European Conference on Computer Vision*, pages 694–711. Springer, 2016.

[63] Simon Baker, Daniel Scharstein, JP Lewis, Stefan Roth, Michael J Black, and Richard Szeliski. A database and evaluation methodology for optical flow. *International Journal of Computer Vision*, 92(1):1–31, 2011.

[64] Robert Keys. Cubic convolution interpolation for digital image processing. *IEEE Transactions on Acoustics, Speech, and Signal Processing*, 29(6):1153–1160, 1981.

[65] Martín Abadi, Paul Barham, Jianmin Chen, Zhifeng Chen, Andy Davis, Jeffrey Dean, Matthieu Devin, Sanjay Ghemawat, Geoffrey Irving, Michael Isard, et al. Tensorflow: A system for

large-scale machine learning. In *12th {USENIX} Symposium on Operating Systems Design and Implementation ({OSDI} 16)*, pages 265–283, 2016.

[66] Tianfan Xue, Baian Chen, Jiajun Wu, Donglai Wei, and William T Freeman. Video enhancement with task-oriented flow. *International Journal of Computer Vision*, 127(8):1106–1125, 2019.

[67] Zhou Wang, Alan C Bovik, Hamid R Sheikh, and Eero P Simoncelli. Image quality assessment: from error visibility to structural similarity. *IEEE Transactions on Image Processing*, 13(4):600–612, 2004.

www.ingramcontent.com/pod-product-compliance
Lightning Source LLC
LaVergne TN
LVHW041715190726
843493LV00007B/2095